MULTIPLE INTEGRALS

G. L. Cohen
Department of Applied Mathematics
The New South Wales Institute of Technology

G. BELL & SONS LTD · LONDON · 1974

First published 1974

Australia
John Cochrane Pty. Ltd, Port Melbourne, Vic.

New Zealand
Book Reps (New Zealand) Ltd, 46 Lake Road, Northcote, Auckland

India
Orient Longman Ltd,
Calcutta, Bombay, Madras and New Delhi

Canada
Clarke, Irwin & Co. Ltd, Toronto

East Africa
J. E. Budds, P.O. Box 4536, Nairobi

West Africa
Thos. Nelson (Nigeria) Ltd, P.O. Box 336, Apapa, Lagos

South and Central Africa
Book Promotions (Pty) Ltd, 311 Sanlam Centre,
Main Road, Wynberg, Cape Province

ISBN 0 7135 1814 6

Text set in 10/11 pt. IBM Press Roman, printed by photolithography, and bound in Great Britain at The Pitman Press, Bath

Preface

This book is written for what is probably the majority of students presently studying mathematics at a tertiary level. The motivation is this: while more and more people are being actively encouraged to undertake tertiary education and being given the opportunity, the range of available text books, particularly beyond those for the first calculus course, is predominantly directed towards the minority of especially competent students who are able to cope with the terminology, abstraction and rigour which only serve to confuse and possibly defeat the average student. The unfortunate aspect of this statement, if accepted, is that the desired level of competence imposed by most institutions on most students is somewhat less than that of the text books they prescribe.

Therefore the approach in the following pages is painstakingly explanatory and illustrative. Analogies with the ordinary definite integral have been emphasised and the common, if often unrecognised, notational difficulties of subscripts and Greek letters, to take two examples, have been minimised. In line with this way of thinking, rigour has of course been played down, but no statement, other than basic problems of the existence of multiple integrals, has been left unproved or at least unqualified. It is hoped that the level of rigour that has been maintained is sufficient for most pass degree students and those on higher diploma and comparable courses.

A knowledge of differentiation (including partial differentiation) and integration is assumed. Elementary use is made of determinants and three dimensional vectors.

Applications of multiple integrals have been delayed until the final chapter, but may be brought forward if and when required. No mention is made in this book of line and surface integrals and associated theorems such as the divergence theorem since it is felt that these are better handled using techniques of vector analysis; but one of the applications included concerns gravitational potential to at least indicate the wider uses of multiple integrals beyond the standard applications of centre of mass and moments of inertia. An application to statistics has also been included.

The author would like to express his appreciation to members of staff and students, too many to name, at the New South Wales Institute of Technology for their suggestions and assistance, particularly in the preparation of the exercises and their solutions. Special thanks are due to Mrs Helen Melville for typing the manuscript.

May 1973 G.L.C.

Contents

1 The Definite Integral Reviewed

The definite integral

Although we shall assume a working knowledge of the theory of integration of functions of a single variable, that is, of the symbol

$$\int_a^b f(x)\,dx,$$

and methods of evaluation for various functions f(x), it will be convenient to recall here some basic background. This will set the style and notation to be followed in subsequent chapters.

The ordinary definite integral is usually constructed in roughly the following manner.

Let f(x) be a continuous function defined in the closed interval $[a, b]$. We partition $[a, b]$ into n subintervals by choosing points $x_0, x_1, x_2, \ldots, x_n$ so that $a = x_0 < x_1 < x_2 < \ldots < x_{n-1} < x_n = b$. Fix on any one of these subintervals and let its length be denoted by Δx. Let x' be some point within that subinterval and consider the product

$$f(x')\,\Delta x.$$

In the usual geometrical interpretation (Fig. 1), where we suppose that $f(x) > 0$ in $[a, b]$, we can see that this product represents the area of the shaded rectangle.

By forming the sum

$$\sum_{\Delta x} f(x')\,\Delta x$$

of such products over all the available subintervals, we form an approximation to the area under the curve $y = f(x)$ between ordinates at $x = a$

and $x = b$. This approximation can be made more precise by increasing the number of subintervals and we are naturally inclined to say that the limit

$$\lim \sum_{\Delta x} f(x')\,\Delta x$$

represents the mentioned area. By 'lim' here we thus mean the limit as $n \to \infty$, though requiring at the same time (and this is not an equivalent statement) that the lengths of the subintervals all tend to zero.

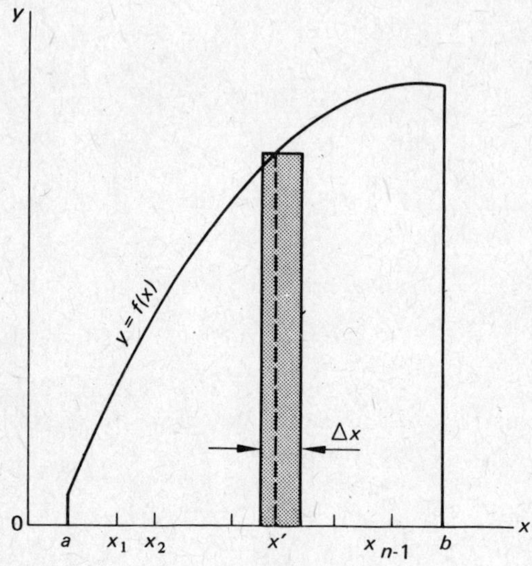

Fig. 1

If this limit exists, it is called the definite integral of $f(x)$ on $[a, b]$ and is denoted by

$$\int_a^b f(x)\,dx.$$

That is,

$$\int_a^b f(x)\,dx = \lim \sum_{\Delta x} f(x')\,\Delta x.$$

It should be noted that when we say 'if this limit exists', the value of such a limit must not depend on the way in which we originally partitioned the interval $[a, b]$, nor on the way we chose the point x'. If it is known that this is so, then we are entitled to choose a particular mode of partitioning and a value of x' which simplify the evaluation of the limit. This is the basis of integration by first principles, examples of which are given in most introductory calculus texts. The problem of determining classes of functions $f(x)$ for which we are assured such limits exist is a difficult one. We will not spend any time on this problem, or on corresponding problems that will arise when we consider double and triple integrals.

As has been said, methods of evaluation of definite integrals are assumed. For example:

(i) the use of substitutions:

$$\int_0^1 \frac{\tan^{-1} x}{1+x^2}\,dx = \int_0^{\pi/4} u\,du \qquad [u = \tan^{-1} x]$$

$$= \left[\frac{1}{2}u^2\right]_0^{\pi/4}$$

$$= \frac{\pi^2}{32};$$

(ii) the use of integration by parts:

$$\int_0^{\pi/3} x \sin x\,dx = -\int_0^{\pi/3} x\,d\cos x$$

$$= -[x\cos x]_0^{\pi/3} + \int_0^{\pi/3} \cos x\,dx$$

$$= -\frac{\pi}{6} + [\sin x]_0^{\pi/3}$$

$$= \frac{3\sqrt{3}-\pi}{6};$$

(iii) the use of partial fractions:

$$\int_1^3 \frac{dx}{x(x^2+1)} = \int_1^3 \left(\frac{1}{x} - \frac{x}{x^2+1}\right) dx$$

$$= [\log x - \tfrac{1}{2}\log(x^2+1)]_1^3$$

$$= \log 3 - \tfrac{1}{2}\log 10 + \tfrac{1}{2}\log 2$$

$$= \log \tfrac{3}{5}\sqrt{5}.$$

A new notation

The notation

$$\int_a^b f(x)\,dx$$

is certainly the most common notation for an ordinary definite integral; but it is not necessarily the most useful one. The same integral could just as well be written

$$\int_I f(x)\,dx,$$

where I is understood to stand for the interval $[a, b]$ and the integration is over this interval.

The advantage in this notation is that I need not stand for just one interval. It could stand for some combination of intervals or any other precisely defined set of points.

In terms of this notation, we now state the fundamental tools used in the evaluation of definite integrals. Their proofs all follow from our definition.

(i) If k is any constant,

$$\int_I kf(x)\,dx = k\int_I f(x)\,dx;$$

(ii) $$\int_I [f(x) + g(x)]\,dx = \int_I f(x)\,dx + \int_I g(x)\,dx;$$

(iii) $$\int_I f(x)\,dx = \int_{I_1} f(x)\,dx + \int_{I_2} f(x)\,dx,$$

where I is expressible as a union of intervals I_1 and I_2 which have no sub-interval in common;

(iv) $$\int_I f(x)\,dx = \int_I f(y)\,dy = \int_I f(t)\,dt = \int_I f(\xi)\,d\xi,$$

that is, the 'x' that we usually use is only a dummy variable and may be replaced by any other symbol.

Notice how, in Example 1 below, we make specific use of these rules.

Their importance should not be underestimated: they demonstrate the linear character of the operator $\int_I$. Linearity is a notion fundamental to most of modern mathematics.

Example 1. Evaluate:

$$\int_I (2x^3 + 1)\,dx$$

where I is the union of intervals $[0, \frac{1}{2}] \cup [1, 3]$.

Solution: Let I_1, I_2 denote the intervals $[0, \frac{1}{2}]$, $[1, 3]$, respectively. Then:

$$\begin{aligned}
\int_{I_1} (2x^3 + 1)\,dx &= \int_{I_1} 2x^3\,dx + \int_{I_1} 1\,dx \\
&= 2\int_{I_1} x^3\,dx + \int_{I_1} dx \\
&= 2\int_0^{1/2} x^3\,dx + \int_0^{1/2} dx \\
&= 2\left[\frac{1}{4}x^4\right]_0^{1/2} + \left[x\right]_0^{1/2} \\
&= \frac{17}{32},
\end{aligned}$$

and, now abbreviating the method in the customary manner,

$$\begin{aligned}
\int_{I_2} (2x^3 + 1)\,dx &= \int_1^3 (2x^3 + 1)\,dx \\
&= \left[\frac{1}{2}x^4 + x\right]_1^3 \\
&= 42.
\end{aligned}$$

Hence,

$$\begin{aligned}
\int_I (2x^3 + 1)\,dx &= \int_{I_1} (2x^3 + 1)\,dx + \int_{I_2} (2x^3 + 1)\,dx \\
&= 42\tfrac{17}{32}.
\end{aligned}$$

Example 2. Evaluate:

$$\int_I (2x^3 + 1)\,dx$$

where I is the interval of the x-axis intercepted by the parabola $y = x^2 - 4x + 3$.

Solution: Before the integral can be transformed to the more familiar form, we must determine where the parabola intercepts the x-axis. This occurs when $y = 0$; that is, when

$$x^2 - 4x + 3 = 0,$$

i.e.

$$(x - 1)(x - 3) = 0,$$

so $x = 1$ and $x = 3$ are the intercepts on the x-axis.

Hence I is the interval $[1, 3]$ and

$$\int_I (2x^3 + 1)\,dx = 42$$

as we calculated in *Example 1.*

Note carefully in this example that the parabola is only there to give us the range of integration. It has no connection whatsoever with the function $2x^3 + 1$ which is to be integrated, other than giving us the range of integration.

EXERCISES 1a

Evaluate the following definite integrals:

(i) $\int_I (3x^2 - x)\,dx$, where I is $[-1, 0] \cup [2, 3]$.

(ii) $\int_I \frac{\log x}{x}\,dx$, where I is $[1, 2] \cup [4, 5]$.

(iii) $\int_I \sin^2\theta\,d\theta$, where I is $[0, \pi/2] \cup [\pi, 2\pi]$.

(iv) $\int_I \log x \, dx$, where I is the interval intercepted on the x-axis by the parabola $y = x^2 - 6x + 8$.

(v) $\int_I \frac{x+2}{x+4} \, dx$, where I is the interval intercepted on the x-axis by the ellipse $\frac{(x-1)^2}{4} + y^2 = 1$.

Trigonometric reduction formulae

There is one particular technique of integration which it is worth revising here in a little more detail. We will find it necessary many times in the coming chapters to evaluate integrals such as:

$$\int_0^{\pi/2} \sin^7 x \, dx, \int_{-\pi/2}^{\pi/2} \cos^8 x \, dx, \int_0^{\pi/2} \sin^7 x \cos^8 x \, dx.$$

Through the use of reduction formulae, such integrals can be handled very easily indeed.

For convenience, in the following we will assume that n and m are non-negative integers. This is the only case of interest to us.

(i) Put

$$I_n = \int_0^{\pi/2} \sin^n x \, dx.$$

Using integration by parts, we obtain, when $n \geqslant 2$,

$$\int \sin^n x \, dx = \int \sin^{n-1} x \sin x \, dx = -\int \sin^{n-1} x \, d \cos x$$

$$= -\cos x \sin^{n-1} x + \int (n-1) \sin^{n-2} x \cos^2 x \, dx$$

$$= -\cos x \sin^{n-1} x + (n-1) \int \sin^{n-2} x \, dx$$

$$- (n-1) \int \sin^n x \, dx,$$

on writing $\cos^2 x = 1 - \sin^2 x$. Hence,

$$n\int \sin^n x \, dx = -\cos x \sin^{n-1} x + (n-1)\int \sin^{n-2} x \, dx$$

so that

$$\int \sin^n x \, dx = \frac{-1}{n}\cos x \sin^{n-1} x + \frac{n-1}{n}\int \sin^{n-2} x \, dx.$$

With the limits 0, $\pi/2$, this says,

$$\int_0^{\pi/2} \sin^n x \, dx = \left[-\frac{1}{n}\cos x \sin^{n-1} x\right]_0^{\pi/2} + \frac{n-1}{n}\int_0^{\pi/2} \sin^{n-2} x \, dx,$$

or simply,

$$I_n = \frac{n-1}{n} I_{n-2}, \quad n \geqslant 2.$$

This result is particularly convenient when n is large, for by repeated application of it the integral may be expressed in terms of I_1 (if n is odd) or I_0 (if n is even), and

$$I_1 = \int_0^{\pi/2} \sin x \, dx = 1,$$

$$I_0 = \int_0^{\pi/2} dx = \pi/2.$$

For example,

$$\text{(a)} \int_0^{\pi/2} \sin^6 x \, dx = I_6$$

$$= \frac{6-1}{6} I_{6-2}$$

$$= \frac{5}{6}\frac{3}{4}\frac{1}{2} I_0$$

$$= \frac{5\pi}{32}.$$

(b) $\int_0^{\pi/2} \sin^5 x \,\mathrm{d}x = I_5$

$$= \frac{4}{5} I_3$$

$$= \frac{4}{5}\frac{2}{3} I_1$$

$$= \frac{8}{15}.$$

Notice that the same formula holds when the limits 0, $\pi/2$ are replaced by any multiples of $\pi/2$, but that the values of I_1 and I_0 may change.

(ii) Put $y = (\pi/2) - x$ in

$$\int_0^{\pi/2} \sin^n x \,\mathrm{d}x.$$

We have $\mathrm{d}y = -\mathrm{d}x$ so that

$$\int_0^{\pi/2} \sin^n x \,\mathrm{d}x = \int_{\pi/2}^{0} \sin^n \left(\frac{\pi}{2} - y\right) (-\mathrm{d}y)$$

$$= \int_0^{\pi/2} \cos^n y \,\mathrm{d}y,$$

and thus precisely the same reduction formula as developed in (i) for $\int_0^{\pi/2} \sin^n x \,\mathrm{d}x$ is valid also for $\int_0^{\pi/2} \cos^n x \,\mathrm{d}x$.

(iii) Put

$$I_{m,n} = \int_0^{\pi/2} \sin^m x \cos^n x \,\mathrm{d}x.$$

It may be shown, with considerable labour, that

$$I_{m,n} = \frac{m-1}{m+n} I_{m-2,n}, \quad m \geqslant 2, \quad n \geqslant 0,$$

or

$$I_{m,n} = \frac{n-1}{m+n} I_{m,n-2}, \quad m \geqslant 0, \quad n \geqslant 2.$$

Again, these are particularly useful when m and n are large. By repeated application of one or other of these formulae, perhaps in combination with the formulae of (i) or (ii), the integral may be expressed in terms of

$$I_{0,0} = \pi/2,$$

or

$$I_{0,1} = I_{1,0} = 1,$$

or

$$I_{1,1} = \tfrac{1}{2}.$$

Benefit is often made of the results

$$I_{m,1} = \frac{1}{m+1}, \qquad m \geqslant 0,$$

$$I_{1,n} = \frac{1}{n+1}, \qquad n \geqslant 0,$$

which are easily proved.

Again, we note that the basic reduction formulae are unchanged,if $0, \pi/2$ are replaced by any other multiples of $\pi/2$, but then the values of $I_{0,0}$, $I_{1,0}, I_{0,1}, I_{1,1}, I_{m,1}, I_{1,n}$ must be checked.

For example,

(a) $$\int_0^{\pi/2} \sin^6 x \cos^5 x \, dx = I_{6,5}$$

$$= \frac{5-1}{6+5} I_{6,3}$$

$$= \frac{4}{11}\frac{2}{9} I_{6,1}$$

$$= \frac{4}{11}\frac{2}{9}\frac{1}{7}$$

$$= \frac{8}{693}.$$

(b) $$\int_0^{\pi} \sin^8 x \cos^6 x \, dx = I_{8,6}$$

$$= \frac{5}{14} I_{8,4}$$

$$= \frac{5}{14}\frac{3}{12}\frac{1}{10} I_{8,0}$$

$$= \frac{5}{14}\frac{3}{12}\frac{1}{10}\frac{7}{8}I_{6,0}$$

$$= \frac{5}{14}\frac{3}{12}\frac{1}{10}\frac{7}{8}\frac{5}{6}\frac{3}{4}\frac{1}{2}I_{0,0}$$

and

$$I_{0,0} = \int_0^\pi dx = \pi.$$

Hence the integral is $5\pi/2^{11}$.

(c) $$\int_0^{\pi/6} \sin^3 3x \cos^7 3x\, dx = \frac{1}{3}\int_0^{\pi/2} \sin^3 y \cos^7 y\, dy \; [y = 3x]$$

$$= \frac{1}{3} I_{3,7}$$

$$= \frac{1}{3}\frac{2}{10} I_{1,7}$$

$$= \frac{1}{3}\frac{2}{10}\frac{1}{8}$$

$$= \frac{1}{120}.$$

With practice, these computations may be carried out in almost a single step.

There are of course many other types of integrals (not necessarily trigonometric ones) that can be handled by similar methods. We shall only need those considered above.

EXERCISES 1b

Find:

(i) $\displaystyle\int_0^{\pi/2} \sin^9 x\, dx,$

(ii) $\displaystyle\int_0^{\pi/2} \cos^8 x\, dx,$

(iii) $\int_0^{\pi} \sin^{10} x \, dx$,

(iv) $\int_0^{\pi/2} \sin^7 x \cos^6 x \, dx$,

(v) $\int_{-\pi}^{\pi} \sin^6 x \cos^4 x \, dx$,

(vi) $\int_0^{\pi/2} \cos^4 x \sin^7 x \, dx$,

(vii) $\int_{-\pi/4}^{\pi/4} \sin^8 x \cos^8 x \, dx$,

(viii) $\int_0^{\pi/4} \sin^5 2x \cos^7 2x \, dx$.

2
Double Integrals

Definition of the double integral

In developing the definition of the double integral, a close watch should be kept on the way the development parallels that of the ordinary definite integral.

Let $f(x, y)$ be a continuous function of the two independent variables x and y, defined in some closed region R of the xy-plane. We partition R into n subregions in some simple fashion as indicated (Fig. 2). Fix on any one of

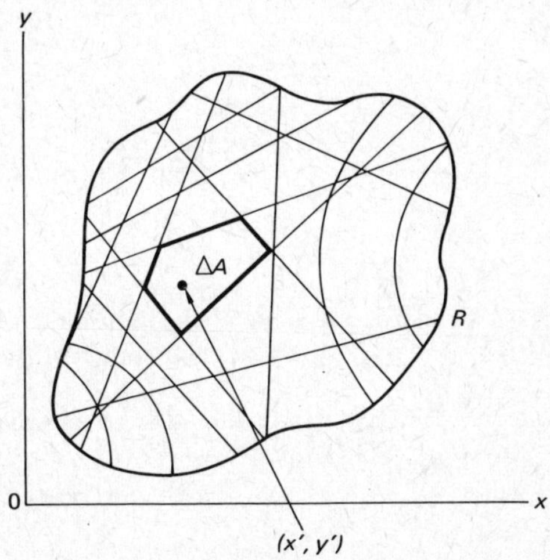

Fig. 2

these subregions and let its area be denoted by ΔA. Let (x', y') be some point within that subregion and consider the product

$$f(x', y')\,\Delta A.$$

Geometrically, we may think of $z = f(x, y)$ as a surface lying over the region R so that this product represents the volume of a column of base area ΔA extending up to the surface. By forming the sum

$$\sum_{\Delta A} f(x', y')\, \Delta A$$

of such products over all the available subregions, we form an approximation to the volume under the surface $z = f(x, y)$ and over R. Reasonably, the more subregions R consists of, the better this approximation will be. We therefore imagine the number n of subregions to be taken indefinitely large so that we may then identify the limit,

$$\lim \sum_{\Delta A} f(x', y')\, \Delta A,$$

as giving the required volume. This should not be understood as a limit in the usual sense: it is somewhat more involved. Although it is a limit 'as $n \to \infty$', it is also necessary that $\Delta A \to 0$ for all possible ΔAs; this may be summarised by requiring only that $(\Delta A)_{\mathbf{max}}$ (the biggest of the ΔAs) tend to zero. Further, when $(\Delta A)_{\mathbf{max}} \to 0$, each subregion must become a point and not, say, a line. This is shown in Fig. 3, in which (a) is

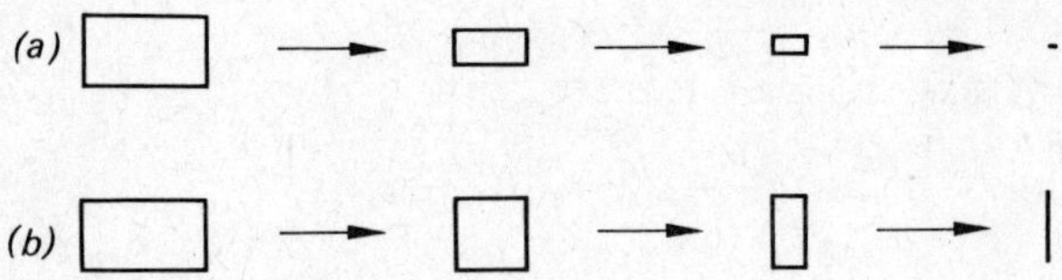

Fig. 3

allowed and (b) is not. Finally, the value of the limit must not depend on the mode of partitioning R or on the particular choice of the point (x', y').

If, subject to all this, the limit exists, then it is called the double integral of $f(x, y)$ over R and is denoted by

$$\iint_R f(x, y)\, dA.$$

That is,

$$\iint_R f(x, y)\, dA = \lim \sum_{\Delta A} f(x', y')\, dA.$$

We now make some more general remarks on this definition.

(i) The notation

$$\iint_R f(x, y)\, dA$$

for a double integral is strikingly close to the notation

$$\int_I f(x)\, dx$$

introduced for the ordinary definite integral. In all respects, it is a natural two-dimensional extension of the notation in the one-dimensional case. In particular, where the notation $\int_I$ implies that definite integrals may be evaluated over more than just the usual interval $[a, b]$, so here $\iint_R$ implies that double integrals exist for quite arbitrary, though well-defined, regions in the xy-plane.

(ii) It should be realised that the definitions of the double integral and the definite integral are quite independent of the geometrical notions which we have introduced. These offer good insight into the way the definitions developed but have no relation to the actual final statement of the definitions. When we come to develop definitions for multiple integrals of functions of more than two independent variables, it is clear that we are very quickly going to run out of geometrical notions to lean on.

(iii) The usual development of integration deals first with indefinite integrals, or antiderivatives. No mention was made of indefinite integrals in Chapter 1 because there is no corresponding concept for multiple integrals. Every double integral (and multiple integral generally) must have associated with it a region of integration.

(iv) In the particular case when $f(x, y) \equiv 1$, we see from the way the definition developed that

$$\iint_R 1\, dA = \iint_R dA = \text{area of } R.$$

(v) Except for the special case in (iv), we have in fact not offered any methods for the evaluation of double integrals. This is our next task.

(vi) We note finally that double integrals satisfy three important properties of linearity, corresponding to those we stated earlier for definite integrals:

(a) If k is any constant,

$$\iint_R k\mathrm{f}(x,y)\,\mathrm{d}A = k\iint_R \mathrm{f}(x,y)\,\mathrm{d}A;$$

(b) $$\iint_R [\mathrm{f}(x,y)+\mathrm{g}(x,y)]\,\mathrm{d}A = \iint_R \mathrm{f}(x,y)\,\mathrm{d}A + \iint_R \mathrm{g}(x,y)\,\mathrm{d}A;$$

(c) $$\iint_R \mathrm{f}(x,y)\,\mathrm{d}A = \iint_{R_1} \mathrm{f}(x,y)\,\mathrm{d}A + \iint_{R_2} \mathrm{f}(x,y)\,\mathrm{d}A,$$

when R is expressible as a union of regions R_1 and R_2 which have at most only boundary points in common.

These properties have obvious extensions (for example, in (b), to a sum of more than two functions).

Repeated integrals

The evaluation of definite integrals in the form $\int_I \mathrm{f}(x)\,\mathrm{d}x$ required first their transformation into the more common form $\int_a^b \mathrm{f}(x)\,\mathrm{d}x$ or a combination of such forms. Similarly, to evaluate double integrals we now derive a method by which they may be transformed into a combination of ordinary definite integrals.

Suppose the region R is convex; that is, it is such that any straight line cuts its boundary in at most two points. For such a region we may speak of 'lower' and 'upper' sides. Let the lower side be given by some function $y = y_1(x)$ and the upper side by some function $y = y_2(x)$ (Fig. 4).

The definition of the double integral implies that we may partition R in any suitable manner and the most convenient would certainly seem to be into rectangles formed by lines parallel to the two axes. Let a typical rectangle so formed have sides Δx and Δy so that for our subregion $\Delta A = \Delta x\,\Delta y$.

We choose our point (x', y') and may now write:

$$\iint_R \mathrm{f}(x,y)\,\mathrm{d}A = \lim \sum_{\Delta A} \mathrm{f}(x',y')\,\Delta y\,\Delta x.$$

We carry out our summation over the various ΔAs in a systematic fashion by fixing a Δx, such as the one in Fig. 4, and considering the rectangles in the column of R lying over it: that is, sum over the different Δys for this fixed Δx. Then move to a different Δx, sum the Δys, move to a different Δx, . . . until the whole summation is completed. If $(\Delta A)_{max} \to 0$ then we must have both $(\Delta x)_{max} \to 0$ and $(\Delta y)_{max} \to 0$ so that we can express the whole process by the equation

$$\lim_{(\Delta A)_{max} \to 0} \sum_{\Delta A} f(x', y')\, \Delta y\, \Delta x$$

$$= \lim_{(\Delta x)_{max} \to 0} \sum_{\Delta x} \left\{ \lim_{(\Delta y)_{max} \to 0} \sum_{\Delta y} f(x', y')\, \Delta y \right\} \Delta x.$$

Look at the inside limit on the right-hand side and compare it with the definition of the definite integral in Chapter 1. With x' fixed, this is

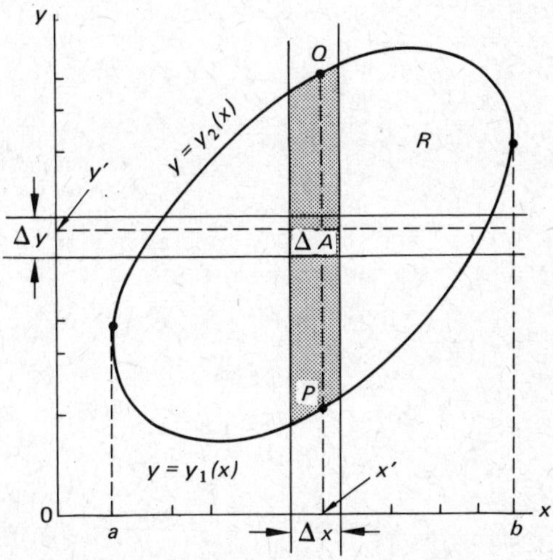

Fig. 4

precisely the integral of the function $f(x', y)$ between the points P and Q in Fig 4. That is,

$$\lim_{(\Delta y)_{max} \to 0} \sum_{\Delta y} f(x', y')\, \Delta y = \int_{y_1(x')}^{y_2(x')} f(x', y)\, dy$$

$$= F(x').$$

This integral will be a function of x' alone, which we have denoted by $F(x')$.

If R is contained between ordinates at $x = a$ and $x = b$ then we now have

$$\lim_{(\Delta A)_{\max} \to 0} \sum_{\Delta A} f(x', y')\, \Delta y\, \Delta x = \lim_{(\Delta x)_{\max} \to 0} \sum_{\Delta x} F(x')\, \Delta x$$

$$= \int_a^b F(x)\, dx,$$

since again we have precisely the definition of a definite integral. Putting all this together, we can write

$$\iint_R f(x, y)\, dA = \int_a^b \left\{ \int_{y_1(x)}^{y_2(x)} f(x, y)\, dy \right\} dx.$$

The double integral has been transformed into a repeated integral (also called an iterated integral): we perform first an integration with respect to y (holding x fixed) and then an integration with respect to x.

We usually write,

$$\iint_R f(x, y)\, dA = \int_a^b \int_{y_1(x)}^{y_2(x)} f(x, y)\, dy\, dx,$$

in which it is understood that the inside integration is to be performed first. The repeated integral is sometimes written

$$\int_a^b dx \int_{y_1(x)}^{y_2(x)} f(x, y)\, dy,$$

and alternative notations for the double integral are:

$$\int_R f(x, y)\, dA \quad \text{and} \quad \iint_R f(x, y)\, dy\, dx.$$

Although the transformation from a double integral into a repeated integral has been shown above for convex regions, the extension to regions which are not convex is generally straightforward and suggests itself for any particular region.

In the following examples, the first few deal with the evaluation of repeated integrals. In each case we show how to obtain the region of integration for the corresponding double integral from the limits on the integrals in the repeated integral. The latter examples deal with double integrals directly. Often the most difficult aspect in the evaluation of a double integral is the determining of the limits on the corresponding repeated integral.

Example 1

$$\int_1^3 \int_2^3 (x^2 - 2xy + 2y^2)\,dy\,dx = \int_1^3 \left[x^2y - xy^2 + \frac{2}{3}y^3\right]_{y=2}^{y=3} dx$$

$$= \int_1^3 \left(x^2 - 5x + \frac{38}{3}\right) dx$$

$$= \left[\frac{1}{3}x^3 - \frac{5}{2}x^2 + \frac{38}{3}x\right]_1^3$$

$$= 14.$$

The integration with respect to y is between the limits $y = 2$ and $y = 3$; the integration with respect to x is between the limits $x = 1$ and $x = 3$.

Hence the corresponding double integral will have a region of integration determined by the inequalities

$$1 \leqslant x \leqslant 3 \quad \text{and} \quad 2 \leqslant y \leqslant 3.$$

This is the rectangle indicated in Fig. 5.If we had been asked to evaluate

$$\iint_R (x^2 - 2xy + 2y^2)\,dA,$$

where R is the above rectangle, we would have first had to transform into the repeated integral of this example.

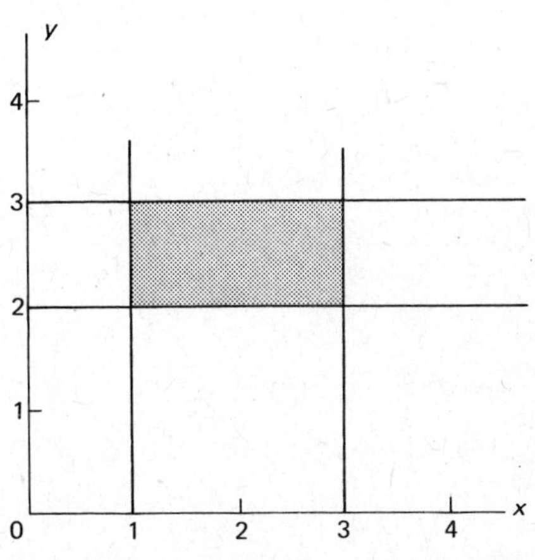

Fig. 5

Example 2. Consider:

$$\int_0^3 \int_{-1}^2 e^{x-2y}\, dy\, dx.$$

Any repeated integral of this form, that is,

$$\int_a^b \int_c^d f(x)\,g(y)\, dy\, dx,$$

where the limits of integration on both integrals are constants and the integrand is expressible as a product of a function of x and a function of y, is called separable, since we may write

$$\int_a^b \int_c^d f(x)\,g(y)\, dy\, dx = \int_a^b f(x)\, dx \cdot \int_c^d g(y)\, dy.$$

This is easily seen from the definition of the repeated integral. Hence,

$$\int_0^3 \int_{-1}^2 e^{x-2y}\, dy\, dx = \int_0^3 e^x\, dx \cdot \int_{-1}^2 e^{-2y}\, dy$$

$$= \Big[e^x\Big]_0^3 \cdot -\tfrac{1}{2}\Big[e^{-2y}\Big]_{-1}^2$$

$$= \tfrac{1}{2}(e^5 - e^2 - e^{-1} + e^{-4}).$$

The region of integration is determined by the inequalities

$$0 \leqslant x \leqslant 3 \quad \text{and} \quad -1 \leqslant y \leqslant 2.$$

It is the square shown in Fig. 6.

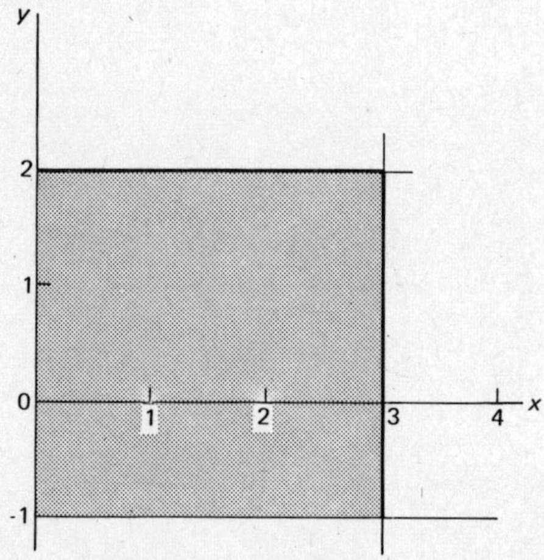

Fig. 6

Example 3

$$\int_0^1 \int_x^{2x} x^2 e^y \, dy \, dx = \int_0^1 x^2 \Big[e^y \Big]_{y=x}^{y=2x} dx$$

$$= \int_0^1 x^2 (e^{2x} - e^x) \, dx$$

$$= \frac{1}{4} e^2 - e + \frac{7}{4},$$

using integration by parts. The region of integration in this case is determined by the inequalities

$$0 \leqslant x \leqslant 1 \quad \text{and} \quad x \leqslant y \leqslant 2x.$$

It is the triangle shown in Fig. 7.

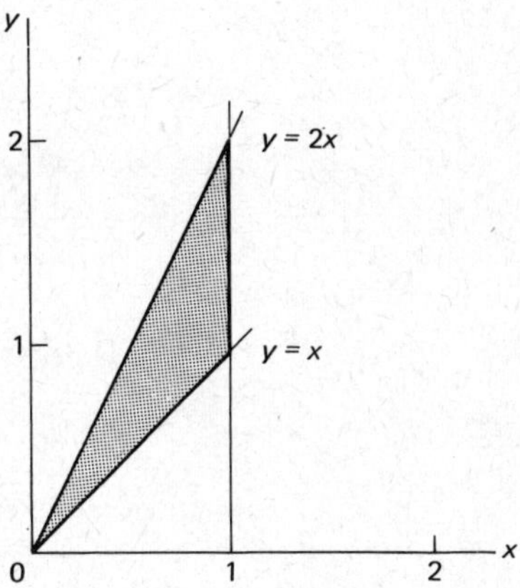

Fig. 7

Example 4

$$\int_0^2 \int_{\sqrt{x}}^{\sqrt{(6-x^2)}} (x+y) \, dy \, dx = \int_0^2 \Big[xy + \tfrac{1}{2} y^2 \Big]_{y=\sqrt{x}}^{y=\sqrt{(6-x^2)}} dx$$

$$= \int_0^2 \Big\{ x\sqrt{(6-x^2)} + \tfrac{1}{2}(6-x^2) - x\sqrt{x} - \tfrac{1}{2}x \Big\} dx$$

$$= \Big[\tfrac{1}{3}(6-x^2)^{3/2} + 3x - \tfrac{1}{6}x^3 - \tfrac{2}{5}x^{5/2} - \tfrac{1}{4}x^2 \Big]_0^2$$

$$= \frac{11}{3} - \frac{34}{15}\sqrt{2} + 2\sqrt{6}.$$

The region of integration is determined by the inequalities

$$0 \leqslant x \leqslant 2 \quad \text{and} \quad \sqrt{x} \leqslant y \leqslant \sqrt{(6 - x^2)}.$$

Thus it lies between ordinates at $x = 0$ and $x = 2$, above the parabola $x = y^2$ and below the circle $x^2 + y^2 = 6$. See Fig. 8.

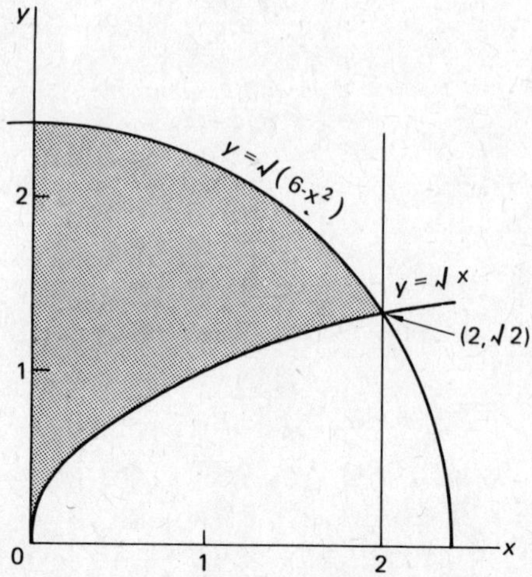

Fig. 8

We state here, as we did in *Example 1*, that if we had been asked to find

$$\iint_R (x + y)\, \mathrm{d}A,$$

where R is the region shown in Fig. 8, the double integral would have to be first transformed into the repeated integral of this example. Notice that this region is not convex, but that this would not give rise to any special difficulties.

Example 5

$$\int_{-1}^{1} \int_{x^2}^{2-x^2} \mathrm{d}y\, \mathrm{d}x = \int_{-1}^{1} \Big[y \Big]_{x^2}^{2-x^2} \mathrm{d}x$$

$$= \int_{-1}^{1} (2 - 2x^2)\, \mathrm{d}x$$

$$= 4\Big[x - \tfrac{1}{3}x^3 \Big]_0^1$$

$$= \frac{8}{3}.$$

The region of integration, determined by the inequalities

$$-1 \leqslant x \leqslant 1 \quad \text{and} \quad x^2 \leqslant y \leqslant 2 - x^2,$$

is shown in Fig. 9. Recall remark (iv) on page 15: we have in fact found the area of the region between the parabolas $y = x^2$ and $y = 2 - x^2$.

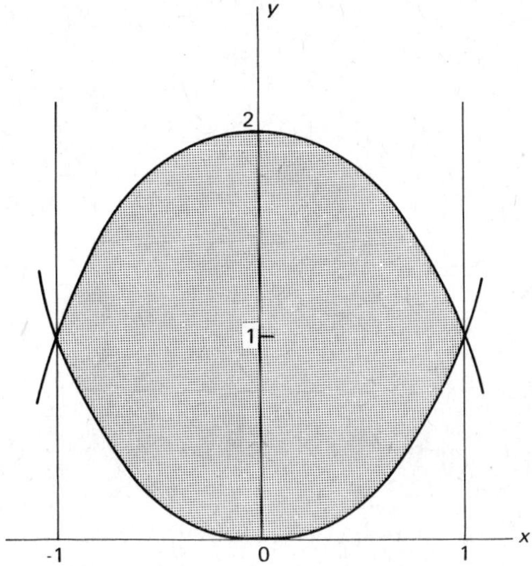

Fig. 9

Example 6. Evaluate:

$$\iint_R xy \, \mathrm{d}A,$$

where R is the region bounded by the lines $y = 0$, $x = 2a$ and the parabola $x^2 = 4ay$.

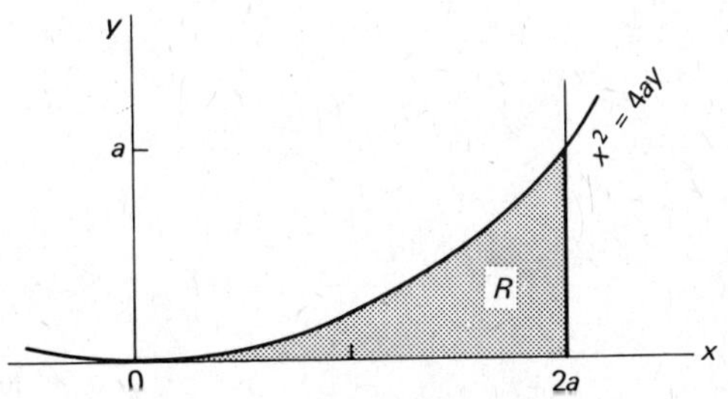

Fig. 10

Solution: This is a double integral. It is evaluated by transforming it into a repeated integral. The limits on the integrals are found from the region R, shown in Fig. 10. We see that R is defined by the inequalities

$$0 \leqslant x \leqslant 2a, \quad 0 \leqslant y \leqslant x^2/4a.$$

Hence,

$$\iint_R xy \, dA = \int_0^{2a} \int_0^{x^2/4a} xy \, dy \, dx$$

$$= \int_0^{2a} x\left[\tfrac{1}{2}y^2\right]_0^{x^2/4a} dx$$

$$= \frac{1}{32a^2}\int_0^{2a} x^5 \, dx$$

$$= \frac{1}{3}a^4.$$

Example 7. Find the area contained between the cubic $y = x^3$ and the parabola $x = y^2$.

Solution: As with all questions on multiple integrals it is a great advantage if an accurate diagram can be quickly drawn. The region referred to here is shown in Fig. 11. We see that it is in fact defined by the inequalities

$$0 \leqslant x \leqslant 1, \quad x^3 \leqslant y \leqslant \sqrt{x}.$$

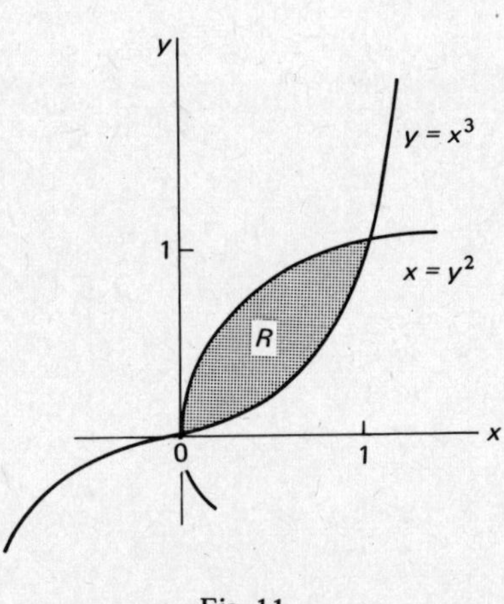

Fig. 11

Denoting this region by R, we thus want

$$\iint_R \mathrm{d}A = \int_0^1 \int_{x^3}^{\sqrt{x}} \mathrm{d}y\,\mathrm{d}x$$

$$= \int_0^1 (\sqrt{x} - x^3)\,\mathrm{d}x$$

$$= \frac{5}{12}.$$

EXERCISES 2a

(1) In each of the following, sketch the region defined by the given inequalities:

(i) $-2 \leqslant x \leqslant 3, \quad 1 \leqslant y \leqslant 3;$

(ii) $0 \leqslant x \leqslant 1, \quad 0 \leqslant y \leqslant x;$

(iii) $0 \leqslant x \leqslant 1, \quad 0 \leqslant y \leqslant x^2;$

(iv) $0 \leqslant x \leqslant 1, \quad 0 \leqslant y \leqslant x^3;$

(v) $0 \leqslant x \leqslant 1, \quad x \leqslant y \leqslant 1;$

(vi) $-2 \leqslant x \leqslant 2, \quad 0 \leqslant y \leqslant 2 + x;$

(vii) $0 \leqslant x \leqslant 3\sqrt{3}, \quad \frac{1}{9}x^2 \leqslant y \leqslant \sqrt{(36 - x^2)};$

(viii) $-a \leqslant x \leqslant a, \quad \frac{1}{2}\sqrt{(a^2 - x^2)} \leqslant y \leqslant \sqrt{(a^2 - x^2)}.$

(2) Evaluate the following repeated integrals and in each case sketch the region of integration:

(i) $\displaystyle\int_0^2 \int_0^1 (x - y)^2\,\mathrm{d}y\,\mathrm{d}x,$

(ii) $\displaystyle\int_0^2 \int_0^x (x - y)^2\,\mathrm{d}y\,\mathrm{d}x,$

(iii) $\displaystyle\int_0^1 \int_x^{\sqrt{x}} (x^2 + y^2)\,\mathrm{d}y\,\mathrm{d}x,$

(iv) $\displaystyle\int_0^1 \int_0^{\sqrt{(1-x^2)}} \frac{xy\,\mathrm{d}y\,\mathrm{d}x}{\sqrt{(1 - y^2)}},$

(v) $\int_0^1 \int_{\sqrt{x}}^{2-\sqrt{x}} xy \, dy \, dx,$

(vi) $\int_{-a}^{a} \int_{\frac{1}{2}\sqrt{(a^2-x^2)}}^{\sqrt{(a^2-x^2)}} 2x^2 y \, dy \, dx.$

(3) Find:

(i) $\iint_R x \, dA,$ (ii) $\iint_R y \, dA,$ (iii) $\iint_R (x+y) \, dA,$

where R is the triangle bounded by the lines $x = 0$, $y = 0$ and $x + y = 1$.

(4) Find:

(i) $\iint_R 2y \, dA,$ (ii) $\iint_R \frac{dA}{\sqrt{(1-y^2)}},$

where R is the region between $y = \sin x$ and $y = 1$ for which $0 \leqslant x \leqslant \pi/2$.

(5) Evaluate the following double integrals over the region stated:

(i) $\iint x^2 y^3 \, dA$ over the rectangle bounded by the lines $x = 0$, $x = 2a$, $y = 0$, $y = 2b$;

(ii) $\iint e^{ax+by} \, dA$ over the triangle bounded by the lines $x = 0$, $y = 0$, $ax + by = 1$ $(a > 0, b > 0)$.

(iii) $\iint x \, dA$ over the smaller segment of the circle $x^2 + y^2 = 4$ cut off by the line $x = 1$.

(iv) $\iint 2y \, dA$ over the region between the ellipse $x^2 + 4y^2 = a^2$ and the circle $x^2 + y^2 = a^2$ in the upper half-plane.

(v) $\iint xy \, dA$ over the region common to the two circles $x^2 + y^2 = 2x$ and $x^2 + y^2 = 2y$.

Transformation to polar coordinates

When, earlier in this chapter, we transformed double integrals into repeated integrals, we did so by considering a particular partitioning of R into rectangles by lines parallel to the axes.

If the integrand is given as a function $g(r, \theta)$ of polar coordinates r and θ, then it is natural to think of a partitioning of R by circles r = constant and rays θ = constant. A more usual application of the result we are about to arrive at involves the transformation of problems initially given in terms of Cartesian coordinates into problems in terms of polar coordinates. When R is easily describable by polar coordinates, any function $f(x, y)$ may be transformed into a function $g(r, \theta)$ by means of the equations

$$x = r\cos\theta,$$
$$y = r\sin\theta,$$

and the resulting integrations may be far simpler by virtue of the simpler description of R.

Suppose then that R is partitioned as in Fig. 12. A typical 'rectangle'

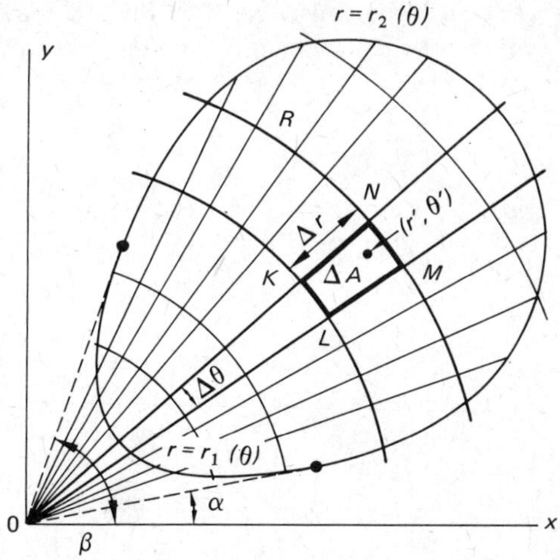

Fig. 12

$KLMN$ in R will have sides $KL = r'\Delta\theta$ and $LM = \Delta r$, so that approximately

$$\Delta A = r'\Delta\theta\,\Delta r,$$

where (r', θ') is some point in our rectangle. Hence,

$$\iint_R g(r, \theta)\,\mathrm{d}A = \lim \sum_{\Delta A} g(r', \theta')\,r'\Delta r\Delta\theta.$$

If we sum over the ΔAs by fixing on $\Delta\theta$, summing over the corresponding Δrs, moving to a new $\Delta\theta$, etc. and since both $(\Delta r)_{\max} \to 0$ and $(\Delta\theta)_{\max} \to 0$, we can write:

$$\lim_{(\Delta A)_{\max} \to 0} \sum_{\Delta A} g(r', \theta') r' \Delta r \Delta\theta$$

$$= \lim_{(\Delta\theta)_{\max} \to 0} \sum_{\Delta\theta} \left\{ \lim_{(\Delta r)_{\max} \to 0} \sum_{\Delta r} g(r', \theta') r' \Delta r \right\} \Delta\theta$$

$$= \lim_{(\Delta\theta)_{\max} \to 0} \sum_{\Delta\theta} G(\theta') \Delta\theta,$$

where

$$G(\theta') = \int_{r_1(\theta')}^{r_2(\theta')} g(r, \theta') r \, dr,$$

$r = r_1(\theta)$ and $r = r_2(\theta)$ being the equations of the 'inner' and 'outer' boundaries of R (which we again assume to be convex at this stage). Finally, if R is contained between rays at $\theta = \alpha$ and $\theta = \beta$,

$$\iint_R g(r, \theta) \, dA = \int_\alpha^\beta G(\theta) \, d\theta$$

$$= \int_\alpha^\beta \left\{ \int_{r_1(\theta)}^{r_2(\theta)} g(r, \theta) r \, dr \right\} d\theta,$$

so our required result is:

$$\iint_R g(r, \theta) \, dA = \int_\alpha^\beta \int_{r_1(\theta)}^{r_2(\theta)} g(r, \theta) r \, dr \, d\theta.$$

Example 1

$$\int_0^{\pi/2} \int_0^{2a \sin\theta} r^2 \sin^2\theta \, dr \, d\theta$$

$$= \int_0^{\pi/2} \sin^2\theta \left[\tfrac{1}{3} r^3 \right]_{r=0}^{r=2a \sin\theta} d\theta$$

$$= \frac{8a^3}{3} \int_0^{\pi/2} \sin^5\theta \, d\theta$$

$$= \frac{8a^3}{3} \frac{4}{5} \frac{2}{3} 1$$

$$= \frac{64a^3}{45} .$$

This is a straight repeated integral. The region of integration is given by the inequalities:

$$0 \leqslant \theta \leqslant \pi/2, \qquad 0 \leqslant r \leqslant 2a \sin\theta,$$

see Fig. 13. Notice our use of the reduction formula derived in Chapter 1.

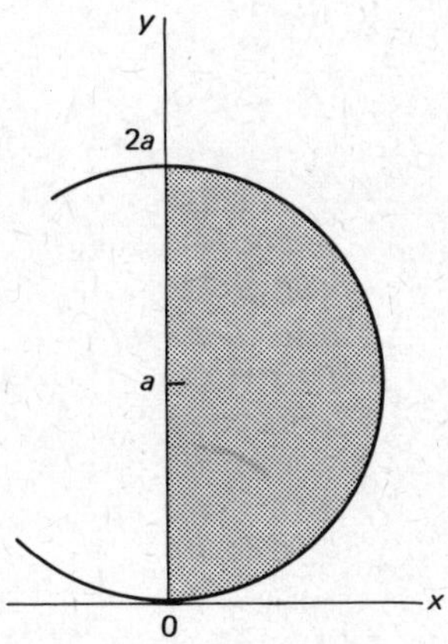

Fig. 13

Example 2. Evaluate:

$$\iint_R y^2 \, dA$$

where R is the region outside the circle $x^2 + y^2 = ax$ and inside the circle $x^2 + y^2 = 2ax$.

Solution: R is the region shown in Fig. 14. It is readily expressible using polar coordinates by the inequalities

$$-\pi/2 \leqslant \theta \leqslant \pi/2, \qquad a \cos\theta \leqslant r \leqslant 2a \cos\theta$$

(since $x^2 + y^2 = ax$ is $r^2 = ar \cos\theta$, etc.). Converting the integrand y^2 to

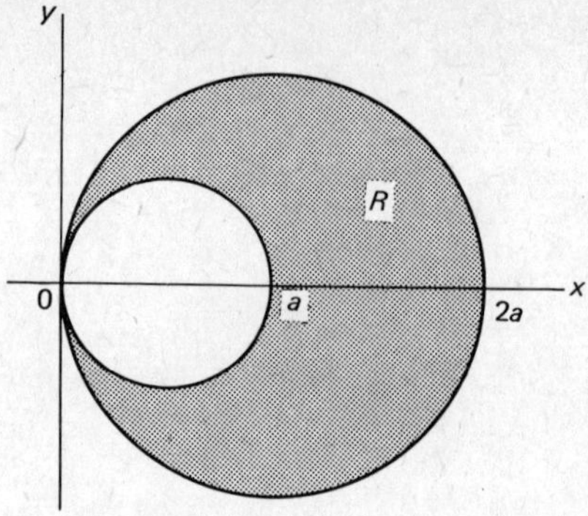

Fig. 14

polar coordinates we obtain $r^2 \sin^2\theta$, so

$$\iint_R y^2\,\mathrm{d}A = \int_{-\pi/2}^{\pi/2} \int_{a\cos\theta}^{2a\cos\theta} r^2 \sin^2\theta \cdot r\,\mathrm{d}r\,\mathrm{d}\theta$$

$$= \int_{-\pi/2}^{\pi/2} \sin^2\theta \left[\tfrac{1}{4} r^4\right]_{a\cos\theta}^{2a\cos\theta} \mathrm{d}\theta$$

$$= \frac{15a^4}{4} \int_{-\pi/2}^{\pi/2} \sin^2\theta \cos^4\theta\,\mathrm{d}\theta$$

$$= \frac{15a^4}{4} \frac{1}{6} \frac{3}{4} \frac{1}{2} \pi$$

$$= \frac{15\pi a^4}{2^6}.$$

EXERCISES 2b

(1) Evaluate the following repeated integrals and sketch the region of integration in each case:

(i) $\displaystyle\int_0^{\pi/2} \int_0^{\sin\theta} r^2 \sin^3\theta\,\mathrm{d}r\,\mathrm{d}\theta,$

(ii) $\displaystyle\int_{\pi/4}^{\pi/2} \int_0^{\operatorname{cosec}\theta} r^3 \cos^2\theta\,\mathrm{d}r\,\mathrm{d}\theta,$

(iii) $\int_0^{\pi/2}\int_{a\cos\theta}^{a} r^2\cos\theta\,\mathrm{d}r\,\mathrm{d}\theta$,

(iv) $\int_0^{\pi}\int_0^{a\sin 2\theta} r\sin^4\theta\,\mathrm{d}r\,\mathrm{d}\theta$.

(2) Use polar coordinates to evaluate the following double integrals over the region stated:

(i) $\iint x^2 y\,\mathrm{d}A$ over that part of the circle $x^2+y^2=a^2$ lying in the first quadrant.

(ii) $\iint (x^2+y^2)\,\mathrm{d}A$ over the whole of the circle $x^2+y^2=a^2$.

(iii) $\iint x^2y^2\,\mathrm{d}A$ over the sector of the circle $x^2+y^2=2$ for which $0 \leqslant y \leqslant x$.

(iv) $\iint \frac{\log(x^2+y^2)}{\sqrt{(x^2+y^2)}}\,\mathrm{d}A$ over the annulus $1 \leqslant x^2+y^2 \leqslant b^2$.

(v) $\iint \sqrt{(a^2-x^2-y^2)}\,\mathrm{d}A$ over the circle $x^2+y^2-ay=0$.

(vi) $\iint \frac{x\,\mathrm{d}A}{(x^2+y^2)^2}$ over the trapezium bounded by the lines $y=0$, $y=x$, $x=\frac{1}{2}$, $x=1$.

(vii) $\iint \frac{x+y}{x^2+y^2}\,\mathrm{d}A$ over the triangle bounded by the lines $x=0$, $y=0$ and $x+y=1$.

(3) Evaluate the following double integrals over the region stated:

(i) $\iint r^5\,\mathrm{d}A$ over the circle $r=a$.

(ii) $\iint \cos\theta \, \mathrm{d}A$ over the region between the circle $r = \cos\theta$ and the cardioid $r = 1 + \cos\theta$, for which $0 \leqslant \theta \leqslant \pi/2$.

(iii) $\iint r \, \mathrm{d}A$ over the region in the first quadrant between the circle $r = \sqrt{3}$ and the rose $r = 2 \sin 2\theta$.

3
Triple Integrals

Definition of the triple integral

The development of a triple integral parallels that of a double integral except that we are now concerned with continuous functions $f(x, y, z)$ of three independent variables x, y and z, so no geometrical representation is available.

Suppose S is a solid region in three-dimensional space and suppose it to be partitioned in some manner into n subregions. Fix on a typical subregion of volume ΔV and let (x', y', z') be an arbitrary point in this subregion. If the sum

$$\sum_{\Delta V} f(x', y', z')\,\Delta V,$$

where the summation extends over all such subregions, has a finite limit as $(\Delta V)_{\mathbf{max}} \to 0$ (requiring also $n \to \infty$), then this limit is called the triple integral of $f(x, y, z)$ over S and is denoted by

$$\iiint_S f(x, y, z)\,dV.$$

That is,

$$\iiint_S f(x, y, z)\,dV = \lim \sum_{\Delta V} f(x', y', z')\,\Delta V.$$

It should be recognised that 'lim' here has the same kind of deeper meaning that we saw when discussing the definition of the double integral.

For the particular case when $f(x, y, z) \equiv 1$, we see by the preceding development that

$$\iiint_S dV = \text{Volume of } S.$$

For this reason, triple integrals are often also called volume integrals.

The evaluation of triple integrals is again effected by transforming them into repeated integrals. The most natural way of partitioning S is by planes parallel to the three coordinate planes. This gives rise to a typical subregion of volume ΔV which is a rectangular prism of base area ΔA and height Δz so that

$$\Delta V = \Delta A \, \Delta z.$$

If both $(\Delta A)_{\max} \to 0$ and $(\Delta z)_{\max} \to 0$, then, as we did with double integrals, we can write the definition of the triple integral as

$$\iiint_S \mathrm{f}(x, y, z) \, \mathrm{d}V$$

$$= \lim_{(\Delta A)_{\max} \to 0} \sum_{\Delta A} \left. \lim_{(\Delta z)_{\max} \to 0} \sum_{\Delta z} \mathrm{f}(x', y', z') \, \Delta z \right\} \Delta A.$$

Assuming for the moment that S is a convex region, so that we can speak of 'lower' and 'upper' surfaces of S given by the functions $z = z_1(x, y)$ and

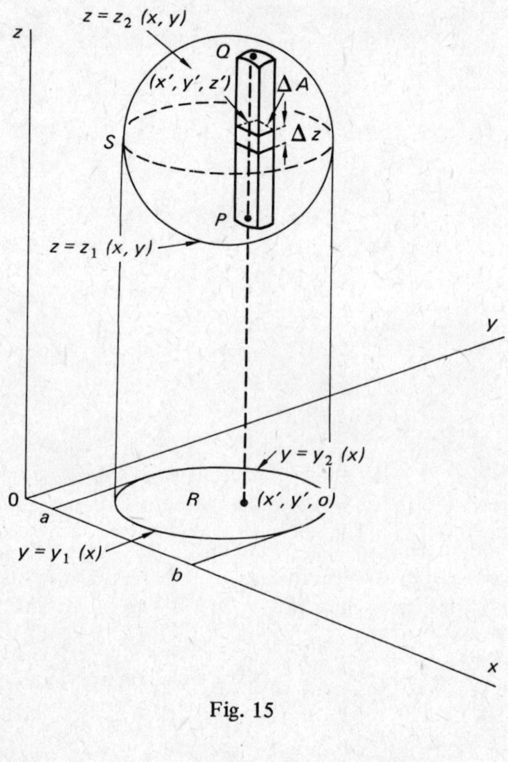

Fig. 15

$z = z_2(x, y)$, respectively, then the inner limit above is simply the integral of $f(x', y', z)$ between the points P and Q in Fig. 15; that is

$$\int_{z_1(x',\,y')}^{z_2(x',\,y')} f(x', y', z)\,dz = F(x', y'),$$

say, a function of x' and y'. We then have

$$\iiint_S f(x, y, z)\,dV = \lim_{(\Delta A)_{\max} \to 0} \sum_{\Delta A} F(x', y')\,\Delta A$$

$$= \iint_R F(x, y)\,dA,$$

by the definition of the double integral, R being the projection of S on the xy-plane. The reduction of double integrals to repeated integrals has been dealt with. We can now write:

$$\iiint_S f(x, y, z)\,dV = \int_a^b \int_{y_1(x)}^{y_2(x)} F(x, y)\,dy\,dx$$

$$= \int_a^b \int_{y_1(x)}^{y_2(x)} \left\{ \int_{z_1(x,\,y)}^{z_2(x,\,y)} f(x, y, z)\,dz \right\} dy\,dx$$

$$= \int_a^b \int_{y_1(x)}^{y_2(x)} \int_{z_1(x,\,y)}^{z_2(x,\,y)} f(x, y, z)\,dz\,dy\,dx,$$

with the usual convention of carrying out the inner integrations first. Thus the triple integral has been transformed to a triply repeated integral.

Notice carefully the form of the repeated integral that corresponds to the triple integral over S. We see that what is required in effect is to describe S (or re-describe it) by inequalities of the form $a \leqslant x \leqslant b$, $y_1(x) \leqslant y \leqslant y_2(x)$, $z_1(x, y) \leqslant z \leqslant z_2(x, y)$: x is given its full allowable range, then y ranges between the functions of x that describe the lower and upper boundaries of R, the projection of S on the xy-plane, and then finally z ranges between the functions of x and y that describe the lower and upper surfaces of S. (In special cases some of the functions mentioned here will be constants.)

Notice next that, as a result, we must perform three integrations. The first is an integration of a function of x, y and z with respect to z; x and y are for the time being treated as constants. The second is an integration of a function of x and y with respect to y; x is still treated as a constant at this stage. Finally there is the integration with respect to x.

It should be mentioned that we are describing here just one of six possible repeated integrals that correspond to a triple integral. It is the one we shall adhere to throughout this section. In the next chapter, we will discuss the other approaches.

If S is not a convex region, then the necessary modifications, if any, are easily dealt with for any particular solid.

Example 1

$$\int_0^2 \int_1^3 \int_2^4 (x+y)z \, dz \, dy \, dx = \int_0^2 \int_1^3 (x+y) \left[\tfrac{1}{2} z^2\right]_2^4 dy \, dx$$

$$= 6 \int_0^2 \int_1^3 (x+y) \, dy \, dx$$

$$= 6 \int_0^2 \left[xy + \tfrac{1}{2} y^2\right]_{y=1}^{y=3} dx$$

$$= 12 \int_0^2 (x+2) \, dx$$

$$= 72.$$

The region of integration, defined by the inequalities:

$$0 \leqslant x \leqslant 2, \quad 1 \leqslant y \leqslant 3, \quad 2 \leqslant z \leqslant 4,$$

is the cube shown in Fig. 16.

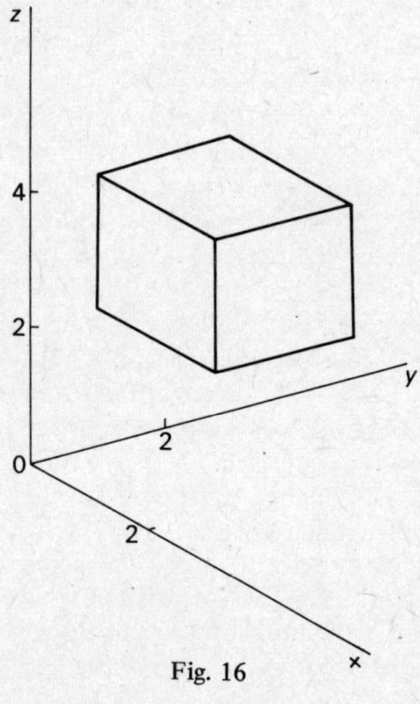

Fig. 16

Example 2

$$\int_0^a \int_0^x \int_0^{\sqrt{(a^2-x^2)}} dz\, dy\, dx$$

$$= \int_0^a \int_0^x \sqrt{(a^2-x^2)}\, dy\, dx$$

$$= \int_0^a x\sqrt{(a^2-x^2)}\, dx$$

$$= -\tfrac{1}{2}\int_0^a (a^2-x^2)^{1/2}\, d(a^2-x^2)$$

$$= -\tfrac{1}{3}\Big[(a^2-x^2)^{3/2}\Big]_0^a$$

$$= \tfrac{1}{3}a^3.$$

The region of integration is defined by the inequalities

$$0 \leqslant x \leqslant a, \quad 0 \leqslant y \leqslant x, \quad 0 \leqslant z \leqslant \sqrt{(a^2-x^2)}.$$

Thus x lies between the planes $x = 0$ and $x = a$, y lies between the planes $y = 0$ and $y = x$ and z lies between the plane $z = 0$ and the cylinder $x^2 + z^2 = a^2$, see Fig. 17.

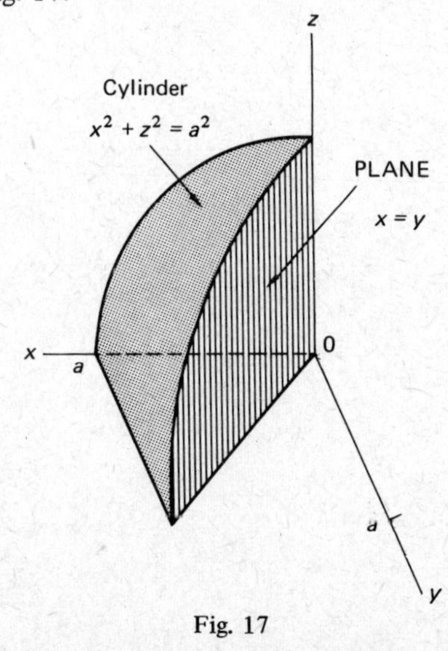

Fig. 17

Since in this example $f(x, y, z) \equiv 1$, we have in fact found the volume of the solid bounded by the planes $y = 0$, $z = 0$, $y = x$ and the cylinder $x^2 + z^2 = a^2$.

Example 3. Evaluate:

$$\iiint z \, dV$$

throughout the tetrahedron bounded by the planes $x = 0$, $y = 0$, $z = 0$ and $x + y + z = 1$.

Solution: The tetrahedron is shown in Fig. 18(*a*). To formulate the triple integral as a repeated integral, we need to specify this region by inequalities. In the preceding theory, the triple integral was expressed in terms of a double integral over the plane region R which is the projection on the xy-plane of our given solid region (see Fig. 18(*b*)). Having hopefully achieved in

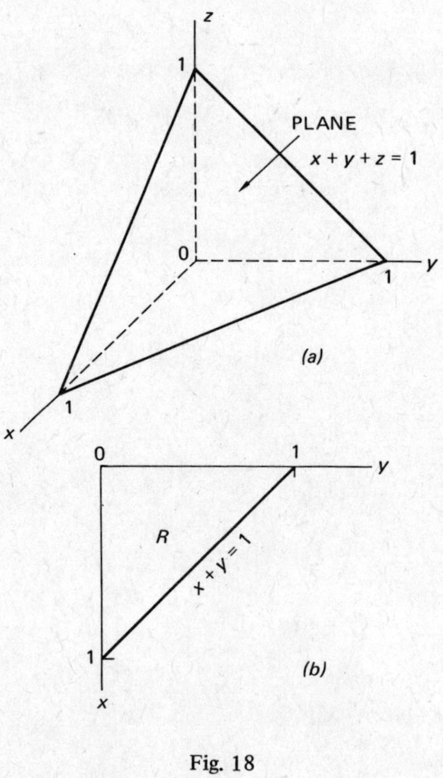

Fig. 18

Chapter 2 a facility for expressing plane regions by inequalities, we see that R is defined by:

$$0 \leqslant x \leqslant 1, \quad 0 \leqslant y \leqslant 1 - x.$$

Reverting to the tetrahedron, z is seen to lie between the planes $z = 0$ and $x + y + z = 1$; that is,

$$0 \leqslant z \leqslant 1 - x - y.$$

Thus,

$$\iiint z \, \mathrm{d}V = \int_0^1 \int_0^{1-x} \int_0^{1-x-y} z \, \mathrm{d}z \, \mathrm{d}y \, \mathrm{d}x$$

$$= \int_0^1 \int_0^{1-x} \left[\tfrac{1}{2} z^2\right]_0^{1-x-y} \mathrm{d}y \, \mathrm{d}x$$

$$= \tfrac{1}{2} \int_0^1 \int_0^{1-x} (1 - x - y)^2 \, \mathrm{d}y \, \mathrm{d}x$$

$$= -\tfrac{1}{6} \int_0^1 \Big[(1 - x - y)^3\Big]_{y=0}^{y=1-x} \mathrm{d}x$$

$$= \tfrac{1}{6} \int_0^1 (1 - x)^3 \, \mathrm{d}x$$

$$= \frac{1}{24}.$$

Example 4. Find the volume common to the paraboloids $z^2 + y^2 = 4ax$ and $z^2 + x^2 = 4ay$.

Solution: Again we seek the projection of the given region on the xy-plane. Denote the solid region by S. Its section with the xy-plane is obtained by putting $z = 0$ in the equations of the paraboloids: we get the parabolas $y^2 = 4ax$ and $x^2 = 4ay$. Hence the required projection is the shaded region in Fig. 19. By symmetry, precisely half of S lies under and over the region marked R between the parabola $x^2 = 4ay$ and the line $y = x$. R is defined by the inequalities

$$0 \leqslant x \leqslant 4a, \quad x^2/4a \leqslant y \leqslant x.$$

Only one paraboloid, $z^2 + x^2 = 4ay$, lies under and over R, and in equal parts. Hence a quarter of the entire region S may be given by

$$0 \leqslant x \leqslant 4a, \quad x^2/4a \leqslant y \leqslant x, \quad 0 \leqslant z \leqslant \sqrt{(4ay - x^2)}.$$

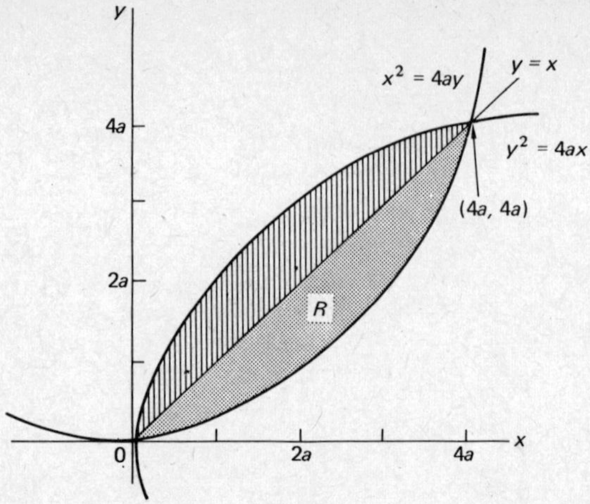

Fig. 19

Thus,

$$\iiint_S \mathrm{d}V = 4\int_0^{4a}\int_{x^2/4a}^{x}\int_0^{\sqrt{(4ay-x^2)}} \mathrm{d}z\,\mathrm{d}y\,\mathrm{d}x$$

$$= 4\int_0^{4a}\int_{x^2/4a}^{x} \sqrt{(4ay-x^2)}\,\mathrm{d}y\,\mathrm{d}x$$

$$= \frac{1}{a}\int_0^{4a}\left[\tfrac{2}{3}(4ay-x^2)^{3/2}\right]_{y=x^2/4a}^{y=x}\mathrm{d}x$$

$$= \frac{2}{3a}\int_0^{4a}(4ax-x^2)^{3/2}\,\mathrm{d}x$$

$$= \frac{2}{3a}\int_0^{4a}\left\{4a^2-(x-2a)^2\right\}^{3/2}\mathrm{d}x$$

$$= \frac{2}{3a}\int_{-\pi/2}^{\pi/2} 8a^3\cos^3\theta\,.\,2a\cos\theta\,\mathrm{d}\theta \qquad [x-2a = 2a\sin\theta]$$

$$= \frac{32a^3}{3}\int_{-\pi/2}^{\pi/2}\cos^4\theta\,\mathrm{d}\theta$$

$$= \frac{32a^3}{3}\,\frac{3}{4}\,\frac{1}{2}\,\pi$$

$$= 4\pi a^3.$$

The required volume is $4\pi a^3$.

EXERCISES 3a

(1) Evaluate the following integrals:

(i) $\displaystyle\int_1^2\int_2^3\int_1^3 (x-y+z)\,dz\,dy\,dx$,

(ii) $\displaystyle\int_0^2\int_0^{x^2}\int_1^{xy} xy\,dz\,dy\,dx$,

(iii) $\displaystyle\int_1^2\int_{2-x}^{2+x}\int_0^{x^2+y^2} x\,dz\,dy\,dx$,

(iv) $\displaystyle\int_0^1\int_x^{\sqrt{x}}\int_{1-x-y}^{1+x+y} xy\,dz\,dy\,dx$.

(2) Evaluate the following triple integrals over the stated regions:

(i) $\iiint(x^2+2y-z)\,dV$ over the box bounded by the planes $x=0$, $x=1, y=0, y=1, z=0, z=1$.

(ii) $\iiint(xz+y)\,dV$ over the box bounded by the planes $x=0, x=4$, $y=0, y=2, z=0, z=2$.

(iii) $\iiint ye^z\,dV$ over that part of the solid bounded by the planes $z=0$ and $z=1$ and the cylinder $x^2+y^2=a^2$ which lies in the first octant.

(iv) $\iiint(x^2+y)\,dV$ over the region bounded by the planes $x=0$, $x=2, y=0, y=1, z=0$ and $x+2y-z+5=0$.

(v) $\iiint e^{(x/a)+(y/b)+(z/c)}\,dV$ over the tetrahedron bounded by the planes $x=0, y=0, z=0$ and $(x/a)+(y/b)+(z/c)=1$ $(a>0, b>0, c>0)$.

(vi) $\iiint\sin(x-y-z)\,dV$ over the tetrahedron bounded by the planes $x=\pi/2, y=0, z=0, z=x-y$.

(3) Show that the following solids have the stated volume:

(i) the solid bounded by the surface $az = xy$ and the planes $x = 0, y = 0$, $z = 0$ and $x + y = a$; $a^3/24$.

(ii) the solid lying above the rectangle $-a \leqslant x \leqslant 2a$, $-a \leqslant y \leqslant 2a$, $z = 0$, and below the surface $z = [(x^2 + y^2)/a] + a$; $27a^3$.

(iii) the solid bounded by the cylinder $y^2 = 4ax$ and the planes $x + z = a$, $z = 0$; $16a^3/15$.

(iv) the solid bounded by the parabolic cylinders $y^2 = ax$ and $x^2 = ay$ and the planes $z = 0$ and $x + y + z = 2a$; $11a^3/30$.

(v) the solid bounded outside by the paraboloid $y^2 + z^2 = 4ax$, inside by the cylinder $y^2 = ax$ and cut off by the plane $x = a$; $\frac{1}{3}(4\pi - 3\sqrt{3})a^3$.

Cylindrical coordinates

Many problems involving double integrals were found to be easier to handle by first transforming into a polar coordinate system. For a corresponding reason, we now indicate how to handle triple integrals by considering them transformed into a system of either cylindrical coordinates or, in the next section, spherical coordinates.

In transforming a triple integral into a repeated integral we arrived at one point at the formula

$$\iiint_S f(x, y, z)\,dV = \iint_R F(x, y)\,dA,$$

where

$$F(x, y) = \int_{z_1(x, y)}^{z_2(x, y)} f(x, y, z)\,dz,$$

and proceeded effectively on the assumption that R, the projection of S on the xy-plane, was partitioned by lines parallel to the x and y axes. We could just as well at that stage have used a polar coordinate partitioning of R, as in Fig. 12, to evaluate the above double integral.

If, on putting $x = r\cos\theta$, $y = r\sin\theta$, the functions $f(x, y, z)$, $F(x, y)$, $z_1(x, y)$, $z_2(x, y)$ become $g(r, \theta, z)$, $G(r, \theta)$, $z_3(r, \theta)$, $z_4(r, \theta)$, respectively, then we would have, just as before,

$$\iint_R F(x, y)\,dA = \iint_R G(r, \theta)\,dA$$

$$= \int_{\alpha}^{\beta} \int_{r_1(\theta)}^{r_2(\theta)} G(r, \theta) r \, dr \, d\theta,$$

where

$$G(r, \theta) = \int_{z_3(r, \theta)}^{z_4(r, \theta)} g(r, \theta, z) \, dz.$$

Our required result is thus,

$$\iiint_S f(x, y, z) \, dV = \int_{\alpha}^{\beta} \int_{r_1(\theta)}^{r_2(\theta)} \int_{z_3(r, \theta)}^{z_4(r, \theta)} g(r, \theta, z) r \, dz \, dr \, d\theta.$$

This formula transforms a problem in Cartesian coordinates to one in cylindrical coordinates. The integrand $f(x, y, z)$ must be transformed by the equations usually written

$$x = r \cos \theta,$$
$$y = r \sin \theta,$$
$$z = z,$$

and the region of integration S must be described in terms of the coordinates z, r, θ. Note that '$r \, dz \, dr \, d\theta$' replaces the former '$dz \, dy \, dx$'.

Example. Determine the volume of the region inside both the sphere $x^2 + y^2 + z^2 = 4a^2$ and the cylinder $(x - a)^2 + y^2 = a^2$.

Solution: The projections of the sphere and cylinder on the xy-plane are shown in Fig. 20. The given solid region S say, projects onto the region

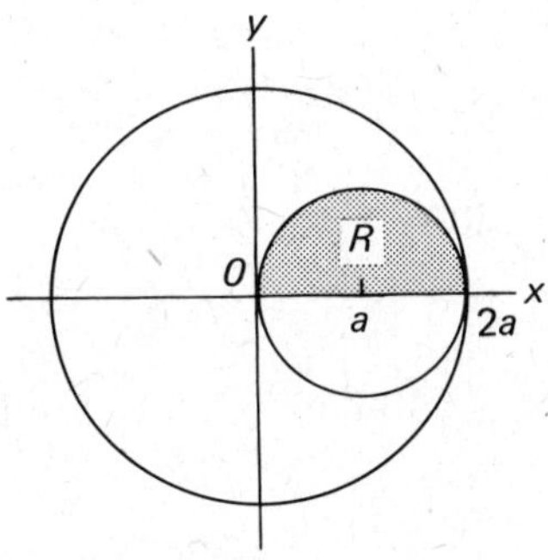

Fig. 20

bounded by the smaller circle. This circle's equation, $x^2 + y^2 = 2ax$, is in polar coordinates $r = 2a \cos \theta$. By symmetry we need consider only that part of S lying over the region R (shaded): this will give us one quarter of the required volume. R is given by the inequalities

$$0 \leqslant \theta \leqslant \pi/2, \qquad 0 \leqslant r \leqslant 2a \cos \theta,$$

and the sphere's equation in cylindrical coordinates is $r^2 + z^2 = 4a^2$, so we take $0 \leqslant z \leqslant \sqrt{(4a^2 - r^2)}$. Hence, the volume of S is

$$\iiint_S \mathrm{d}V = 4\int_0^{\pi/2} \int_0^{2a\cos\theta} \int_0^{\sqrt{(4a^2 - r^2)}} r \,\mathrm{d}z \,\mathrm{d}r \,\mathrm{d}\theta$$

$$= 4\int_0^{\pi/2} \int_0^{2a\cos\theta} r\sqrt{(4a^2 - r^2)} \,\mathrm{d}r \,\mathrm{d}\theta$$

$$= -\frac{4}{3}\int_0^{\pi/2} \left[(4a^2 - r^2)^{3/2}\right]_0^{2a\cos\theta} \mathrm{d}\theta$$

$$= -\frac{4}{3}.\, 8a^3 \int_0^{\pi/2} (\sin^3\theta - 1)\,\mathrm{d}\theta$$

$$= -\frac{32}{3}a^3\left(\frac{2}{3} - \frac{\pi}{2}\right)$$

$$= \frac{16}{9}a^3\,(3\pi - 4).$$

EXERCISES 3b

(1) Use cylindrical coordinates to find the following integrals, where S is the portion of the cylinder $x^2 + y^2 = a^2$ between the planes $z = 0$ and $z = h$.

(i) $\iiint_S \sqrt{(x^2 + y^2)}\,\mathrm{d}V,$

(ii) $\iiint_S x^2 y^2 z\,\mathrm{d}V,$

(iii) $\iiint_S z^2 \sin(x^2 + y^2)\,\mathrm{d}V.$

(2) Find the following integrals, where S is that part of the cylindrical

shell $1 \leqslant x^2 + y^2 \leqslant a^2$ in the first octant between the planes $z = 1$ and $z = h$ $(a > 1, h > 1)$.

(i) $$\iiint_S \frac{xy}{z}\,\mathrm{d}V,$$

(ii) $$\iiint_S \left(\frac{xy}{\sqrt{(x^2+y^2)}}\right)^3 \mathrm{d}V.$$

(3) Find the following integrals, where S is the cylinder $x^2 + y^2 = ay$ cut off by the planes $z = 0$ and $z = h$.

(i) $$\iiint_S (x^2 + y^2)\,\mathrm{d}V,$$

(ii) $$\iiint_S (x^2y^2 + z^2)\,\mathrm{d}V.$$

(4) Find the volume of the solid bounded by the cylinder $x^2 + y^2 = a^2$, the paraboloid $z = b(x^2 + y^2)$, $b > 0$, and the plane $z = 0$.

(5) Show that the volume of the wedge formed by the cylinder $x^2 + y^2 = 2ax$ and the planes $z = mx$, $z = nx$, $n > m$, is $\pi(n - m)a^3$.

(6) Find the volume enclosed by the cylinders $x^2 + y^2 = 2ax$ and $z^2 = 2ax$.

(7) Show that

(i) $$\iiint_S (x^2 + y^2 + z^2)\,\mathrm{d}V = \frac{2^{10}a^5k}{75}\left(1 + \frac{k^2}{3}\right),$$

(ii) $$\iiint_S (x^2y^2 + y^2z^2 + z^2x^2)\,\mathrm{d}V = \frac{2^{13}a^7k}{735}\left(k^2 + \frac{8}{33}\right),$$

where S is the region enclosed by the cylinder $x^2 + y^2 = 2ax$ and the upper and lower sections of the cone $z^2 = k^2(x^2 + y^2)$.

Spherical coordinates

To obtain a transformation law from Cartesian to spherical coordinates, we must revert to the definition of the triple integral and partition our solid region differently: we use planes θ = constant, halfcones ϕ = constant and spheres ρ = constant. Fig. 21(*a*) shows the spherical coordinates (ρ, ϕ, θ) of a point P, together with its Cartesian coordinates (x, y, z) and cylindrical coordinates (r, θ, z).

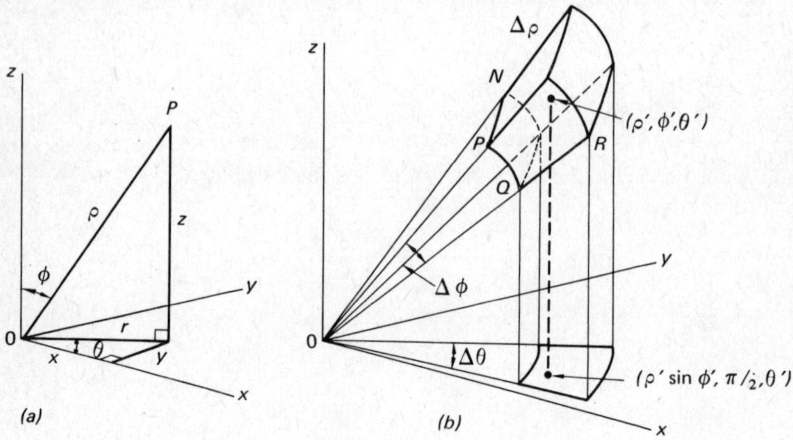

Fig. 21

Such a partitioning leads to the typical volume element ΔV shown in Fig. 21(*b*). It is seen to be an approximately rectangular prism. Let (ρ', ϕ', θ') be some point within the element. The projection of the lowest face of the prism is also shown so that the lengths of the edges can be written down as approximately,

$$PQ = \rho'\Delta\phi, \quad QR = \Delta\rho \text{ and } NP = \rho' \sin \phi'\Delta\theta.$$

Hence,

$$\Delta V = \rho'^2 \sin \phi'\Delta\rho\, \Delta\phi\, \Delta\theta,$$

approximately. Fig. 21(*a*) implies that the function $f(x, y, z)$ may be transformed into a function $g(\rho, \phi, \theta)$ by the equations

$$x = \rho \sin \phi \cos \theta,$$
$$y = \rho \sin \phi \sin \theta,$$
$$z = \rho \cos \phi.$$

We then have

$$\iiint_S f(x, y, z)\, dV = \iiint_S g(\rho, \phi, \theta)\, \rho^2 \sin \phi\, d\rho\, d\phi\, d\theta,$$

where on the right the region S is to be expressed in terms of spherical coordinates.

Note that now '$\rho^2 \sin\phi \, d\rho \, d\phi \, d\theta$' replaces the former '$dz \, dy \, dx$' and that the full allowable range of spherical coordinates is

$$\rho \geqslant 0, \qquad 0 \leqslant \theta < 2\pi, \qquad 0 \leqslant \phi \leqslant \pi.$$

Example 1. Evaluate:

$$\iiint_S (x^2 + y^2 + z^2) \, dV$$

where S is the interior of the sphere $x^2 + y^2 + z^2 = a^2$.

Solution: In spherical coordinates, the integrand $x^2 + y^2 + z^2$ becomes ρ^2 and the sphere has equation $\rho = a$. By symmetry, we need only consider that part of the sphere in the first octant; this is described by the inequalities

$$0 \leqslant \theta \leqslant \pi/2, \qquad 0 \leqslant \phi \leqslant \pi/2, \qquad 0 \leqslant \rho \leqslant a.$$

Hence,

$$\iiint_S (x^2 + y^2 + z^2) \, dV = 8 \int_0^{\pi/2} \int_0^{\pi/2} \int_0^a \rho^2 \,.\, \rho^2 \sin\phi \, d\rho \, d\phi \, d\theta$$

$$= 8 \int_0^{\pi/2} \int_0^{\pi/2} \sin\phi \left[\frac{1}{5}\rho^5\right]_0^a d\phi \, d\theta$$

$$= \frac{8}{5}a^5 \int_0^{\pi/2} d\theta$$

$$= \frac{4}{5}\pi a^5.$$

Example 2. Use spherical coordinates to describe that part of the interior of the cone $x^2 + y^2 = xz$ between the planes $z = 0$ and $z = k$ $(k > 0)$, and evaluate $\iiint x \, z \, dV$ over this region.

Solution: Fig. 22 shows the sloping cone of height k. (Its base is a circle of diameter k.) To describe the region in spherical coordinates, we see first that

$$-\pi/2 \leqslant \theta \leqslant \pi/2.$$

The cone's equation is:

$$\rho^2 \sin^2 \phi (\cos^2 \theta + \sin^2 \theta) = \rho^2 \sin \phi \cos \phi \cos \theta$$

or

$$\tan \phi = \cos \theta .$$

Hence,

$$0 \leqslant \phi \leqslant \tan^{-1} \cos \theta .$$

Finally, the plane has equation $\rho \cos \phi = k$, so $0 \leqslant \rho \leqslant k \sec \phi$.

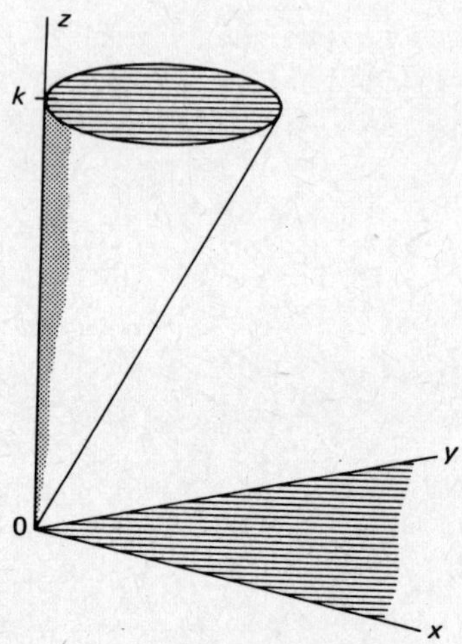

Fig. 22

The required region may thus be described by the inequalities

$$-\pi/2 \leqslant \theta \leqslant \pi/2, \qquad 0 \leqslant \phi \leqslant \tan^{-1} \cos \theta, \quad 0 \leqslant \rho \leqslant k/\cos \phi.$$

The integrand xz becomes $\rho^2 \sin \phi \cos \phi \cos \theta$ in spherical coordinates; so, using the symmetry of the region (and since xz is positive throughout it),

$$\iiint xz \, dV = 2 \int_0^{\pi/2} \int_0^{\tan^{-1} \cos \theta} \int_0^{k/\cos \phi} \rho^2 \sin \phi \cos \phi \cos \theta \, . \, \rho^2 \sin \phi \, d\rho \, d\phi \, d\theta$$

$$= 2 \int_0^{\pi/2} \int_0^{\tan^{-1} \cos\theta} \cos \theta \sin^2 \phi \cos \phi \left[\tfrac{1}{5} \rho^5 \right]_0^{k/\cos \phi} d\phi \, d\theta$$

$$= \tfrac{2}{5} k^5 \int_0^{\pi/2} \cos\theta \int_0^{\tan^{-1}\cos\theta} \tan^2\phi \, d\tan\phi \, d\theta$$

$$= \tfrac{2}{5} k^5 \int_0^{\pi/2} \cos\theta \left[\tfrac{1}{3} \tan^3\phi \right]_0^{\tan^{-1}\cos\theta} d\theta$$

$$= \frac{2}{15} k^5 \int_0^{\pi/2} \cos^4\theta \, d\theta$$

$$= \frac{2}{15} k^5 \, \frac{3}{4} \, \frac{1}{2} \, \frac{\pi}{2}$$

$$= \frac{1}{40} \pi k^5.$$

EXERCISES 3c

(1) Use spherical coordinates to find the following integrals, where S is the part of the sphere $x^2 + y^2 + z^2 \leqslant a^2$ in the first octant:

(i) $$\iiint_S x^2 \, dV,$$

(ii) $$\iiint_S xyz \, dV,$$

(iii) $$\iiint_S \frac{z}{x^2 + y^2 + z^2} dV.$$

(2) Use a triple integral to find the volume of a sphere.

(3) Evaluate:

$$\iiint_S \frac{dV}{(x^2 + y^2 + z^2)^{3/2}},$$

where S is the region between the spheres $x^2 + y^2 + z^2 = a^2$ and $x^2 + y^2 + z^2 = b^2$ $(a > b)$.

(4) Evaluate:

$$\iiint_S (x^2 y^2 + y^2 z^2 + z^2 x^2) \, dV,$$

where S is the interior of the sphere $x^2 + y^2 + z^2 = 1$.

(5) A plane cuts off a cap of height h from a sphere of radius a. If the plane is taken parallel to the xy-plane with the centre of the sphere at the origin of coordinates, show that the volume of the cap is

$$\int_0^{2\pi}\int_0^{\cos^{-1}[1-(h/a)]}\int_{(a-h)/\cos\phi}^{a} \rho^2 \sin\phi \, d\rho \, d\phi \, d\theta .$$

Evaluate this integral.

Use cylindrical coordinates to obtain an alternative expression for the volume of the cap, and check the value of the integral.

(6) Do (7) of Exercises 3b using spherical coordinates.

4

Changing the Order of Integration

It has probably become apparent that, wherever in the past we have converted a multiple integral into a repeated integral, the development has given rise to a certain order in which the integrations are to be carried out and we have stuck to that order in all the examples and exercises.

For example, with double integrals, we always integrated first with respect to y, then with respect to x, or if polar coordinates were used, first with respect to r, then with respect to θ. Triple integrals have been evaluated by integrating in the orders z, y, x or z, r, θ or ρ, ϕ, θ.

In general there is no good reason why we have kept so rigidly to these orders, except that they have been wholly adequate for evaluating the integrals that have arisen so far. But suppose, for example, the integrand were

$$x^2 e^{-y^2}.$$

Integration in our usual order (y, then x) is impossible here: e^{-y^2} has no antiderivative (expressible as an elementary function). If the order were reversed, we would begin by integrating x^2, and this offers no difficulty. The integration with respect to y still remains to be done, but as we shall soon see in an example, the region of integration may be such as to make this possible; certainly, we have at least made a beginning. There are other ways in which, depending on the region of integration, the process of reversing the order of integration in a repeated integral may either permit the integral to be evaluated, when otherwise it would be impossible, or may simplify the evaluation, when otherwise it would be tedious.

The theory may be illustrated by returning to our original double integral. Fig. 23 is much the same as Fig. 4 (p. 17), but we have indicated that R (since it is convex) may be thought of as having 'left' and 'right' sides given by the functions $x = x_1(y)$ and $x = x_2(y)$, respectively, and we have contained R between abscissae at $y = c$ and $y = d$.

Instead of the earlier equation,

$$\lim_{(\Delta A)_{\max} \to 0} \sum_{\Delta A} f(x', y')\, \Delta y\, \Delta x$$

$$= \lim_{(\Delta x)_{\max} \to 0} \sum_{\Delta x} \left\{ \lim_{(\Delta y)_{\max} \to 0} \sum_{\Delta y} f(x', y')\, \Delta y \right\} \Delta x,$$

we could equally well write,

$$\lim_{(\Delta A)_{\max} \to 0} \sum_{\Delta A} f(x', y')\, \Delta y\, \Delta x$$

$$= \lim_{(\Delta y)_{\max} \to 0} \sum_{\Delta y} \left\{ \lim_{(\Delta A')_{\max} \to 0} \sum_{\Delta x} f(x', y')\, \Delta x \right\} \Delta y,$$

the implication here being that Δy is fixed and we sum over the corresponding Δxs before moving to a different Δy, and so on. By the definition of the definite integral, the inner limit on the right here is the integral of $f(x, y')$ between the points S and T in Fig. 23; that is,

$$\int_{x_1(y')}^{x_2(y')} f(x, y')\, dx = G(y'),$$

say. Then,

$$\iint_R f(x, y)\, dA = \lim_{(\Delta y)_{\max} \to 0} \sum_{\Delta y} G(y')\, \Delta y$$

$$= \int_c^d G(y)\, dy$$

$$= \int_c^d \int_{x_1(y)}^{x_2(y)} f(x, y)\, dx\, dy.$$

The double integral has thus been expressed as a repeated integral where the order of integration is the reverse of the former order. It follows that, provided the double integral exists,

$$\int_a^b \int_{y_1(x)}^{y_2(x)} f(x, y)\, dy\, dx = \int_c^d \int_{x_1(y)}^{x_2(y)} f(x, y)\, dx\, dy.$$

It is important to notice that changing the order of integration does not simply mean writing 'dx dy' instead of 'dy dx' or writing

$$`\int_{y_1(x)}^{y_2(x)} \int_a^b `$$

instead of

$$`\int_a^b \int_{y_1(x)}^{y_2(x)} `$$

(unless and only if $y_1(x)$ and $y_2(x)$ are constants). The preceding equation shows that R must be reexpressed by inequalities of the form $c \leqslant y \leqslant d$, $x_1(y) \leqslant x \leqslant x_2(y)$. This time, of course, the inner integration is with respect to x, and y is held constant.

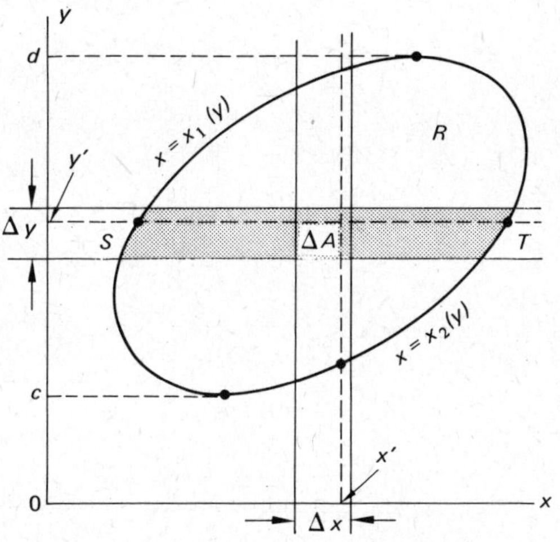

Fig. 23

Similarly we may change the order of integration in triple repeated integrals (there are six possible orders of integration), no matter what coordinate system is being used. Refer to Fig. 15 (p. 34): for Cartesian coordinates we obtain one of the other five orders by integrating over the projection R in the order x, then y. The remaining four orders of integration arise by projecting S on the yz- or zx-planes.

The most convenient order for any particular problem depends on either the integrand or the region of integration, or both, in that problem.

Example 1. Verify:

$$\int_2^3 \int_3^4 (x+y)\,\mathrm{d}y\,\mathrm{d}x = \int_3^4 \int_2^3 (x+y)\,\mathrm{d}x\,\mathrm{d}y.$$

Solution: We have

$$\int_2^3 \int_3^4 (x+y)\,\mathrm{d}y\,\mathrm{d}x = \int_2^3 \left[xy + \tfrac{1}{2}y^2\right]_{y=3}^{y=4} \mathrm{d}x$$

$$= \int_2^3 \left(x + \frac{7}{2}\right)\mathrm{d}x$$

$$= \left[\frac{1}{2}x^2 + \frac{7}{2}x\right]_2^3$$

$$= 6,$$

and

$$\int_3^4 \int_2^3 (x+y)\,\mathrm{d}x\,\mathrm{d}y = \int_3^4 \left[\tfrac{1}{2}x^2 + xy\right]_{x=2}^{x=3} \mathrm{d}y$$

$$= \int_3^4 \left(y + \frac{5}{2}\right)\mathrm{d}y$$

$$= \left[\frac{1}{2}y^2 + \frac{5}{2}y\right]_3^4$$

$$= 6,$$

which verifies the given equation.

In cases like this, and only in such cases, where the region of integration is a rectangle with sides parallel to the axes (that is, where the limits on the integrals are constants), changing the order of integration amounts to a straight swap of the integrations.

Example 2. Evaluate the integral:

$$\int_0^1 \int_x^1 x^2 \mathrm{e}^{-y^2}\,\mathrm{d}y\,\mathrm{d}x.$$

Solution: It is impossible to carry out the integration in the given order. The region of integration (see Fig. 24) is given by the inequalities

$$0 \leqslant x \leqslant 1, \qquad x \leqslant y \leqslant 1.$$

It may equally well be specified by the inequalities

$$0 \leqslant y \leqslant 1, \qquad 0 \leqslant x \leqslant y.$$

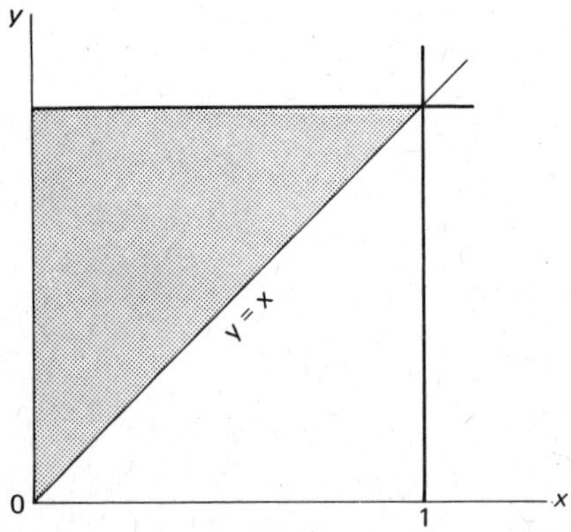

Fig. 24

We thus have

$$\int_0^1 \int_x^1 x^2 e^{-y^2}\, dy\, dx = \int_0^1 \int_0^y x^2 e^{-y^2}\, dx\, dy$$

$$= \int_0^1 e^{-y^2} \left[\tfrac{1}{3}x^3\right]_{x=0}^{x=y} dy$$

$$= \tfrac{1}{3}\int_0^1 y^3 e^{-y^2}\, dy$$

$$= \tfrac{1}{6}\int_0^1 t e^{-t}\, dt \qquad [t = y^2]$$

$$= \tfrac{1}{6}(1 - 2e^{-1}),$$

using integration by parts.

In this example, the region of integration dissolves the apparent impasse caused by having an integrand which possesses no antiderivative.

(In fairness, we mention the following interesting way of evaluating this integral. It avoids the necessity of reversing the order of integration by treating the integral as an ordinary definite integral and integrating by parts. Put

$$f(x) = \int_x^1 e^{-y^2}\, dy.$$

Then,

$$\int_0^1 x^2 f(x)\, dx = \int_0^1 f(x)\, d(\tfrac{1}{3}x^3)$$

$$= \left[\tfrac{1}{3}x^3 f(x)\right]_0^1 - \tfrac{1}{3}\int_0^1 x^3\, d\, f(x)$$

$$= \tfrac{1}{3} f(1) - \tfrac{1}{3}\int_0^1 x^3 f'(x)\, dx.$$

But $f(1) = 0$ and, using the Fundamental Theorem of the Calculus, $f'(x) = -e^{-x^2}$. Hence

$$\int_0^1 x^2 \left(\int_x^1 e^{-y^2}\, dy\right) dx = \tfrac{1}{3}\int_0^1 x^3 e^{-x^2}\, dx,$$

which is one of the steps we arrived at by the preceding method, with x as the dummy variable instead of y.)

Example 3. Evaluate:

$$\iint xy\, dA$$

over the trapezium bounded by the lines $y = 0, y = 1, y = x, 3x + 4y = 12$.

Solution: The region of integration is shown in Fig. 25. We depart from integrating in the usual order, y then x, for we see that we would in fact need to sum three integrals, namely,

$$\int_0^1 \int_0^x xy\, dy\, dx + \int_1^{8/3} \int_0^1 xy\, dy\, dx + \int_{8/3}^4 \int_0^{(12-3x)/4} xy\, dy\, dx.$$

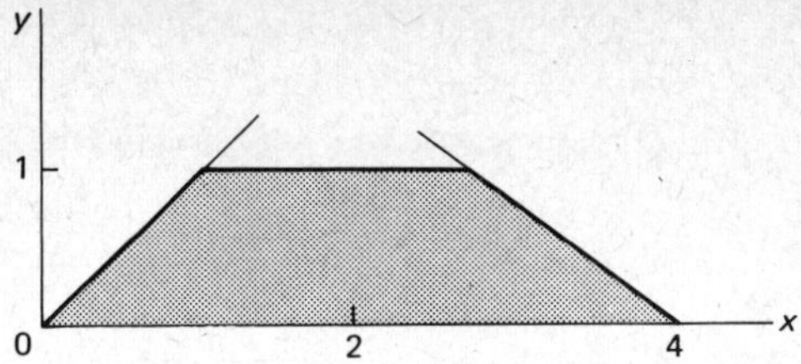

Fig. 25

With the opposite order of integration, there is just one integral:

$$\int_0^1 \int_y^{(12-4y)/3} xy \, dx \, dy = \int_0^1 y\left[\tfrac{1}{2} x^2\right]_y^{(12-4y)/3} dy$$

$$= \frac{1}{18}\int_0^1 (144y - 96y^2 + 7y^3)\, dy$$

$$= \frac{167}{72}.$$

EXERCISES 4

(1) In each of parts (i) to (viii) in Exercises 2a(1), specify the region by giving inequalities for y and x in that order.

(2) In each of parts (i) to (v) in Exercises 2a(2), check the value of the integral by first inverting the order of integration.

(3) Verify that:

(i) $$\int_0^1 \int_y^1 e^{-x^2} \, dx \, dy = \tfrac{1}{2}\left(1 - \frac{1}{e}\right),$$

(ii) $$\int_0^1 \int_x^1 e^{x/y} \, dy \, dx = \tfrac{1}{2}(e - 1),$$

(iii) $$\int_0^1 \int_y^1 \frac{\sin x}{x} \, dx \, dy = 1 - \cos 1,$$

(iv) $$\int_0^1 \int_x^1 y^4 e^{xy^2} \, dy \, dx = \tfrac{1}{3}(e - 2).$$

(4) By first inverting the order of integration, show that

$$\int_{a/2}^{a}\int_{x}^{a}\frac{x\,dy\,dx}{(x^2+y^2)^{3/2}}=\log(2+\sqrt{5})-\log(1+\sqrt{2})-(1/\sqrt{2})\log 2.$$

(5) Prove that

$$\int_0^k\int_0^x f(y)\,dy\,dx=\int_0^k (k-y)\,f(y)\,dy.$$

Use this result to deduce that

$$\int_0^k\int_0^x\int_0^y f(z)\,dz\,dy\,dx=\tfrac{1}{2}\int_0^k (k-z)^2 f(z)\,dz,$$

and hence evaluate

$$\int_0^1\int_0^x\int_0^y \frac{e^z}{(1-z)^2}\,dz\,dy\,dx.$$

(6) Prove that

(i) $$\int_0^1\int_0^{1-\sqrt{(1-x^2)}}\frac{xye^y}{(y-1)^2}\,dy\,dx=\tfrac{1}{2},$$

(ii) $$\int_0^1\int_0^{1-\sqrt{(1-x^2)}}\frac{xy\log(y+1)}{(y-1)^2}\,dy\,dx=\tfrac{1}{8}.$$

(7) Indicate in a diagram the region over which the following repeated integrals are taken:

$$\int_0^{a/\sqrt{2}}\int_0^x \cos\left\{c(x^2+y^2)\right\}dy\,dx$$

$$+\int_{a/\sqrt{2}}^{a}\int_0^{\sqrt{(a^2-x^2)}}\cos\left\{c(x^2+y^2)\right\}dy\,dx.$$

By changing the order of integration, express the sum of these integrals as one repeated integral and evaluate the integral by first transforming into polar coordinates.

(8) Evaluate:

$$\iiint_S \sqrt{(x^2 + y^2 + z^2)}\,\mathrm{d}V,$$

where S is the region bounded by the plane $z = 3$ and the cone $z = \sqrt{(x^2 + y^2)}$, by transforming into cylindrical coordinates and integrating in the order r, z, θ.

5

Change of Variables in Multiple Integrals

In the evaluation of ordinary definite integrals, written in the form $\int_I f(x)\,dx$ say, it is a common practice to convert the integral into a more recognisable form, by changing the variable from x to u through some function $x = x(u)$. The converted integral then has the form $\int_{I'} g(u)\left|\frac{dx}{du}\right| du$, where $g(u) = f[x(u)]$ is the function which $f(x)$ becomes on putting $x = x(u)$ and I' is the interval or set of points to which I is transformed by the substitution.

As a simple but direct application of the method of change of variable stated in this form, consider the integral

$$\int_I \frac{\sin^{-1} x}{\surd(1-x^2)}\,dx,$$

where I is the interval $[0, 1]$. We make the substitution $x = \sin u$. The integrand becomes $u/\cos u$; the derivative dx/du is $\cos u$; corresponding to the interval $[0, 1]$, we have $[0, \pi/2]$. Hence the integral is equal to

$$\int_0^{\pi/2} (u/\cos u) \,.\, \cos u \,.\, du = \int_0^{\pi/2} u\,du$$

$$= \pi^2/8.$$

Our aim in this chapter is to develop a corresponding method for the evaluation of multiple integrals. With definite integrals, a change of variable is normally made with the aim of producing a more recognisable integrand, as has been stated; the interval I is generally transformed to just another interval I'. However, with multiple integrals there is a second motive for changing variables: the region of integration may often be transformed through a substitution into a more straightforward region, thus simplifying in this way the subsequent integrations. This will be seen in examples later on.

A word of caution before we proceed: in this chapter we need some knowledge of partial derivatives, determinants and vectors, so some revision of these notions may be required.

We shall derive the desired rule for double integrals only. The corresponding result, proved in a corresponding manner, will then be stated for triple integrals.

In the integral,

$$\iint_R \mathrm{f}(x, y)\, \mathrm{d}A,$$

we change the variables from x and y to new independent variables u and v by the functions

$$\left.\begin{aligned} x &= x(u, v) \\ y &= y(u, v) \end{aligned}\right\}.$$

For convenience, we are here using (as we did above) the same letters x and y to denote both the functional relationships and the dependent variables, but this should not cause any difficulty.

We shall assume that these equations can be solved to give u and v as functions of x and y:

$$\left.\begin{aligned} u &= u(x, y) \\ v &= v(x, y) \end{aligned}\right\}.$$

This means that to each point in the uv-plane there corresponds a point in the xy-plane, and to each point in the xy-plane there corresponds a point in the uv-plane.

Such equations are said to define a transformation or mapping of the plane.

To the region R in the xy-plane there corresponds a region R' in the uv-plane, and vice versa (see Fig. 26). Suppose the region R' is partitioned by lines parallel to the u- and v-axes. Then we may choose a typical subregion (in the uv-plane) of area $\Delta A'$, say. This is a rectangle and so we may write:

$$\Delta A' = \Delta u\, \Delta v.$$

Denote this rectangle by $K'L'M'N'$, as shown. In the xy-plane there will be a partitioning of R corresponding to that of R' and a corresponding subregion $KLMN$ of area ΔA say. We must relate the two areas ΔA and $\Delta A'$.

NOTE: In the typeface adopted for this book, the italic υ appears as v throughout.

Suppose K' is the point (u, v). Then the other points of the rectangle will be $L'(u + \Delta u, v)$, $M'(u + \Delta u, v + \Delta v)$, $N'(u, v + \Delta v)$. In the xy-plane the corresponding points may thus be written as $K[x(u, v), y(u, v)]$, $L[x(u + \Delta u, v), y(u + \Delta u, v)]$, $M[x(u + \Delta u, v + \Delta v), y(u + \Delta u, v + \Delta v)]$ and $N[x(u, v + \Delta v), y(u, v + \Delta v)]$.

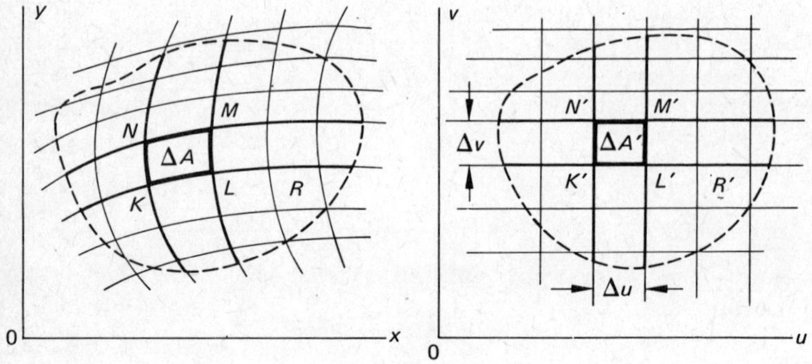

Fig. 26

For a differentiable function $y = f(x)$ of a single variable x, we recall, from the definition of the derivative,

$$f(x + \Delta x) - f(x) = f'(x)\,\Delta x + \epsilon \Delta x,$$

where $\epsilon \to 0$ as $\Delta x \to 0$. If Δx is small, this means that

$$f(x + \Delta x) = f(x) + f'(x)\,\Delta x,$$

approximately. Careful application of this result to our situation allows us to write

$$x(u + \Delta u, v) = x(u, v) + x_u(u, v)\,\Delta u,$$

$$x(u + \Delta u, v + \Delta v) = x(u, v) + x_u(u, v)\,\Delta u + x_v(u, v)\,\Delta v,$$

and so on, as reasonable first approximations, provided Δu and Δv are small. Changing the notation, we may thus write our points as:

$$K(x, y),\ L\left(x + \frac{\partial x}{\partial u}\Delta u,\ y + \frac{\partial y}{\partial u}\Delta u\right),$$

$$M\left(x + \frac{\partial x}{\partial u}\Delta u + \frac{\partial x}{\partial v}\Delta v,\ \ y + \frac{\partial y}{\partial u}\Delta u + \frac{\partial y}{\partial v}\Delta v\right)$$

and

$$N\left(x + \frac{\partial x}{\partial v}\Delta v,\ y + \frac{\partial y}{\partial v}\Delta v\right).$$

Now introduce the vectors

$$KL = \frac{\partial x}{\partial u}\Delta u\ \mathbf{i} + \frac{\partial y}{\partial u}\Delta u\ \mathbf{j},$$

$$KN = \frac{\partial x}{\partial v}\Delta v\ \mathbf{i} + \frac{\partial y}{\partial v}\Delta v\ \mathbf{j},$$

$$KM = \left(\frac{\partial x}{\partial u}\Delta u + \frac{\partial x}{\partial v}\Delta v\right)\mathbf{i} + \left(\frac{\partial y}{\partial u}\Delta u + \frac{\partial y}{\partial v}\Delta v\right)\mathbf{j}.$$

We notice immediately that $KL + KN = KM$, so that the approximations we made above mean effectively that we consider *KLMN* to be a parallelogram. Using a well-known result concerning the cross product of two vectors, we may now determine the area ΔA:

$$\Delta A = |KL \times KN|,$$

and

$$KL \times KN = \begin{vmatrix} \mathbf{i} & \mathbf{j} & \mathbf{k} \\ \frac{\partial x}{\partial u}\Delta u & \frac{\partial y}{\partial u}\Delta u & 0 \\ \frac{\partial x}{\partial v}\Delta v & \frac{\partial y}{\partial v}\Delta v & 0 \end{vmatrix}$$

$$= \left(\frac{\partial x}{\partial u}\frac{\partial y}{\partial v} - \frac{\partial x}{\partial v}\frac{\partial y}{\partial u}\right)\Delta u\ \Delta v\ \mathbf{k}.$$

In parentheses we have the determinant

$$\begin{vmatrix} \frac{\partial x}{\partial u} & \frac{\partial x}{\partial v} \\ \frac{\partial y}{\partial u} & \frac{\partial y}{\partial v} \end{vmatrix},$$

commonly called the Jacobian of x and y with respect to u and v and denoted by

$$\frac{\partial(x, y)}{\partial(u, v)} \quad \text{or} \quad \mathrm{J}\left(\frac{x, y}{u, v}\right).$$

Putting everything together, we have

$$\Delta A = \left|\frac{\partial(x, y)}{\partial(u, v)}\right| \Delta A',$$

giving the required relationship between ΔA and $\Delta A'$. This is of course only an approximate result. Like all such statements, the approximation is improved by taking smaller values for $\Delta A'$. The Jacobian may be looked upon as the magnification factor relating the areas ΔA and $\Delta A'$. It is a function of x and y (or of u and v) which, in the context so far, we may imagine to be evaluated at some point within the rectangle $K'L'M'N'$. Let this be the point (u', v'), and let (x', y') be the corresponding point in $KLMN$.

Suppose $f(x, y)$ becomes $g(u, v)$ on substituting $x = x(u, v)$, $y = y(u, v)$. Then, by definition of the double integral,

$$\iint_R f(x, y)\, dA = \lim_{(\Delta A)_{\max} \to 0} \sum_{\Delta A} f(x', y')\, \Delta A$$

$$= \lim_{(\Delta A')_{\max} \to 0} \sum_{\Delta A'} g(u', v') \left|\frac{\partial(x, y)}{\partial(u, v)}\right|_{u', v'} \Delta A'$$

$$= \iint_{R'} g(u, v) \left|\frac{\partial(x, y)}{\partial(u, v)}\right| dA'.$$

The rule is usually written:

$$\iint_R f(x, y)\, dx\, dy = \iint_{R'} g(u, v) \left|\frac{\partial(x, y)}{\partial(u, v)}\right| du\, dv.$$

Notice that the Jacobian of the transformation to cylindrical coordinates, by the functions $x = r \cos \theta$, $y = r \sin \theta$, is

$$\frac{\partial(x, y)}{\partial(r, \theta)} = \begin{vmatrix} \dfrac{\partial x}{\partial r} & \dfrac{\partial x}{\partial \theta} \\ \dfrac{\partial y}{\partial r} & \dfrac{\partial y}{\partial \theta} \end{vmatrix}$$

$$= \begin{vmatrix} \cos \theta & -r \sin \theta \\ \sin \theta & r \cos \theta \end{vmatrix}$$

$$= r \cos^2 \theta + r \sin^2 \theta$$

$$= r,$$

so that

$$\iint_R f(x, y)\, dx\, dy = \iint_{R'} g(r, \theta)\, r\, dr\, d\theta,$$

as derived in Chapter 2. Polar coordinates are just one particular example of possible alternative coordinate systems.

In a similar manner it may be shown that the triple integral

$$\iiint_S f(x, y, z)\, dV$$

may be transformed through the equations

$$\left.\begin{aligned} x &= x(u, v, w) \\ y &= y(u, v, w) \\ z &= z(u, v, w) \end{aligned}\right\}$$

according to the rule:

$$\iiint_S f(x, y, z)\, dx\, dy\, dz = \iiint_{S'} g(u, v, w) \left|\frac{\partial(x, y, z)}{\partial(u, v, w)}\right| du\, dv\, dw,$$

where S' is the region in uvw-space to which S is transformed, and

$$\frac{\partial(x, y, z)}{\partial(u, v, w)}$$

is the three-dimensional Jacobian:

$$\begin{vmatrix} \dfrac{\partial x}{\partial u} & \dfrac{\partial x}{\partial v} & \dfrac{\partial x}{\partial w} \\[2ex] \dfrac{\partial y}{\partial u} & \dfrac{\partial y}{\partial v} & \dfrac{\partial y}{\partial w} \\[2ex] \dfrac{\partial z}{\partial u} & \dfrac{\partial z}{\partial v} & \dfrac{\partial z}{\partial w} \end{vmatrix}$$

Notice that the Jacobian of the transformation to cylindrical coordinates, defined by the functions $x = r\cos\theta$, $y = r\sin\theta$, $z = z$, is:

$$\frac{\partial(x, y, z)}{\partial(r, \theta, z)} = \begin{vmatrix} \dfrac{\partial x}{\partial r} & \dfrac{\partial x}{\partial \theta} & \dfrac{\partial x}{\partial z} \\[2ex] \dfrac{\partial y}{\partial r} & \dfrac{\partial y}{\partial \theta} & \dfrac{\partial y}{\partial z} \\[2ex] \dfrac{\partial z}{\partial r} & \dfrac{\partial z}{\partial \theta} & \dfrac{\partial z}{\partial z} \end{vmatrix}$$

$$= \begin{vmatrix} \cos\theta & -r\sin\theta & 0 \\ \sin\theta & r\cos\theta & 0 \\ 0 & 0 & 1 \end{vmatrix}$$

$$= r,$$

and that for spherical coordinates, defined by $x = \rho \sin \phi \cos \theta$, $y = \rho \sin \phi \sin \theta$, $z = \rho \cos \phi$, is:

$$\frac{\partial(x, y, z)}{\partial(\rho, \phi, \theta)} = \begin{vmatrix} \frac{\partial x}{\partial \rho} & \frac{\partial x}{\partial \phi} & \frac{\partial x}{\partial \theta} \\ \frac{\partial y}{\partial \rho} & \frac{\partial y}{\partial \phi} & \frac{\partial y}{\partial \theta} \\ \frac{\partial z}{\partial \rho} & \frac{\partial z}{\partial \phi} & \frac{\partial z}{\partial \theta} \end{vmatrix}$$

$$= \begin{vmatrix} \sin \phi \cos \theta & \rho \cos \phi \cos \theta & -\rho \sin \phi \sin \theta \\ \sin \phi \sin \theta & \rho \cos \phi \sin \theta & \rho \sin \phi \cos \theta \\ \cos \phi & -\rho \sin \phi & 0 \end{vmatrix}$$

$$= \cos \phi \,.\, \rho \cos \phi \,.\, \rho \sin \phi \begin{vmatrix} \cos \theta & -\sin \theta \\ \sin \theta & \cos \theta \end{vmatrix}$$

$$+ \rho \sin \phi \,.\, \sin \phi \,.\, \rho \sin \phi \begin{vmatrix} \cos \theta & -\sin \theta \\ \sin \theta & \cos \theta \end{vmatrix}$$

$$= \rho^2 \sin \phi \, (\cos^2 \phi + \sin^2 \phi) \, (\cos^2 \theta + \sin^2 \theta)$$

$$= \rho^2 \sin \phi.$$

These would lead us to the results previously obtained in Chapter 3.

The Jacobian of the transformation $x = x(u, v)$, $y = y(u, v)$ may often be determined by finding the Jacobian of the inverse transformation $u = u(x, y)$, $v = v(x, y)$ and applying the reciprocal theorem:

$$\frac{\partial(x, y)}{\partial(u, v)} \cdot \frac{\partial(u, v)}{\partial(x, y)} = 1.$$

This result may be seen as follows. Exploying the subscript notation for convenience, the left-hand side is:

$$(u_x v_y - u_y v_x)(x_u y_v - x_v y_u)$$

$$= (u_x x_u + u_y y_u)(v_x x_v + v_y y_v) - (u_x x_v + u_y y_v)(v_x x_u + v_y y_u)$$

$$= u_u v_v - u_v v_u$$

$$= 1 \times 1 - 0 \times 0$$

$$= 1,$$

where we have used the chain rule for partial derivatives. If the elements of matrix multiplication are known, the identity used here is not as obscure as it may appear.

The reciprocal theorem suggests that if the transformation $x = x(u, v)$, $y = y(u, v)$ is to be applied to a double integral, then neither of the Jacobians in the theorem can be zero at any point in the region of integration. This is in fact necessary; it is the condition under which the inverse transformation exists.

In general the Jacobian will be a continuous function of x and y so, since it cannot be zero in the region of integration, it must remain always positive or always negative. Notice that its absolute value is required in the rule for change of variables, so that in effect we take it to be always positive.

The reciprocal theorem is easily extended to three-dimensional Jacobians.

Example 1. Find the area of the region bounded by the parabolas $y^2 = x$, $y^2 = 2x$ and the rectangular hyperbolas $xy = 1$, $xy = 2$.

Solution: We seek $\iint_R \mathrm{d}A$ where R is shown in Fig. 27. Under the transformation,

$$u = y^2/x, \qquad v = xy,$$

the region R' in the uv-plane corresponding to R is clearly simply the square bounded by $u = 1$, $u = 2$, $v = 1$ and $v = 2$. This is why this particular transformation was chosen.

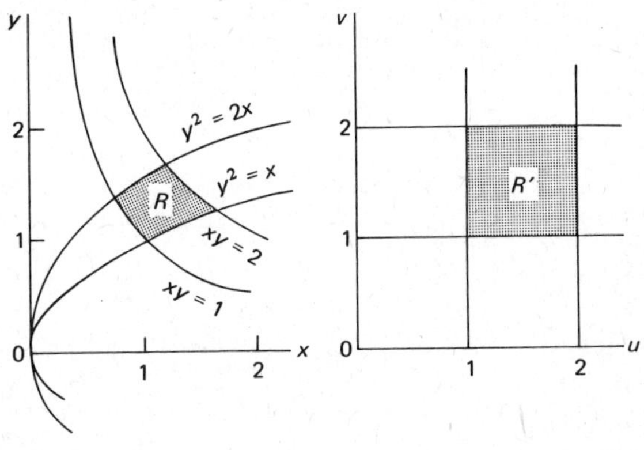

Fig. 27

We need the Jacobian

$$\frac{\partial(x, y)}{\partial(u, v)}.$$

In this case, it is simpler to compute $\partial(u, v)/\partial(x, y)$ and apply the reciprocal theorem:

$$\frac{\partial(u, v)}{\partial(x, y)} = \begin{vmatrix} \dfrac{\partial u}{\partial x} & \dfrac{\partial u}{\partial y} \\ \dfrac{\partial v}{\partial x} & \dfrac{\partial v}{\partial y} \end{vmatrix}$$

$$= \begin{vmatrix} -\dfrac{y^2}{x^2} & \dfrac{2y}{x} \\ y & x \end{vmatrix}$$

$$= -3\frac{y^2}{x}$$

$$= -3u,$$

so

$$\frac{\partial(x, y)}{\partial(u, v)} = \frac{-1}{3u}.$$

Hence,

$$\iint_R dx\, dy = \iint_{R'} \left| \frac{\partial(x, y)}{\partial(u, v)} \right| du\, dv$$

$$= \int_1^2 \int_1^2 \frac{1}{3u}\, du\, dv$$

$$= \tfrac{1}{3} \log 2.$$

Note that if the work had been carried through in Cartesian coordinates, we would have had to obtain the four points of intersection and then to evaluate

$$\int_{2^{-1/3}}^{1} \int_{1/x}^{\sqrt{(2x)}} dy\, dx + \int_{1}^{2^{1/3}} \int_{\sqrt{x}}^{\sqrt{(2x)}} dy\, dx + \int_{2^{1/3}}^{2^{2/3}} \int_{\sqrt{x}}^{2/x} dy\, dx.$$

Example 2. Evaluate:

$$\iint_R x^2 y\, dA$$

where R is described by $(x^2/a^2) + (y^2/b^2) \leqslant 1, y \geqslant 0$.

Solution: If we make the substitution

$$u = x/a, \qquad v = y/b,$$

then R will be replaced by the semicircular region R': $u^2 + v^2 \leqslant 1, v \geqslant 0$.

$$\frac{\partial(x, y)}{\partial(u, v)} = \begin{vmatrix} a & 0 \\ 0 & b \end{vmatrix}$$

$$= ab,$$

so

$$\iint_R x^2 y \, dx \, dy = \iint_{R'} (au)^2 (bv) \left| \frac{\partial(x, y)}{\partial(u, v)} \right| du \, dv$$

$$= a^3 b^2 \iint_{R'} u^2 v \, du \, dv.$$

This integration is best done in polar coordinates. Put $u = r \cos \theta$, $v = r \sin \theta$. Then,

$$\frac{\partial(u, v)}{\partial(r, \theta)} = r$$

and the integral

$$= a^3 b^2 \int_0^\pi \int_0^1 (r \cos \theta)^2 (r \sin \theta) \, r \, dr \, d\theta$$

$$= a^3 b^2 \int_0^1 r^4 \, dr \, . \int_0^\pi \cos^2 \theta \sin \theta \, d\theta$$

$$= \frac{2}{15} a^3 b^2.$$

In Examples 1 and 2 transformations were employed which provided more straightforward regions of integration. In Example 3, the transformation simplifies the integrand as well as the region of integration.

Example 3. By means of the transformation:

$$x = uv, \qquad y = u(1 - v),$$

show that

$$\iint_R (y/x^2) \log (x + y) \, dx \, dy = (2 \log 2 - 1)(1 - \log 2),$$

where R is the trapezium bounded by the lines $x + y = 1, x + y = 2, x = y$ and $y = 0$.

Solution: The inverse transformation is seen to be

$$u = x + y, \qquad v = x/(x + y)$$

so the lines $x + y = 1, x + y = 2, x = y, y = 0$ are mapped respectively onto the lines $u = 1, u = 2, v = \frac{1}{2}, v = 1$. These lines define a rectangle R' in the uv-plane into which R is transformed.

The Jacobian of the transformation is:

$$\frac{\partial(x, y)}{\partial(u, v)} = \begin{vmatrix} v & u \\ 1 - v & -u \end{vmatrix}$$

$$= -u.$$

Hence,

$$\iint_R (y/x^2) \log (x + y) \, dx \, dy = \iint_{R'} \frac{u(1 - v)}{(uv)^2} \log u \left| \frac{\partial(x, y)}{\partial(u, v)} \right| du \, dv$$

$$= \int_{1/2}^{1} \int_{1}^{2} \frac{1 - v}{v^2} \log u \, du \, dv$$

$$= \int_{1/2}^{1} \frac{1 - v}{v^2} \, dv \, . \int_{1}^{2} \log u \, du$$

$$= \left[-\frac{1}{v} - \log v \right]_{1/2}^{1} . \left[u \log u - u \right]_{1}^{2}$$

$$= (1 - \log 2)(2 \log 2 - 1),$$

as required.

Example 4. Find the volume of the solid determined by $0 \leqslant \sqrt{x} + \sqrt{y} + \sqrt{z} \leqslant \sqrt{a}$, where $a > 0$.

Solution: We require $\iiint_S dV$, where S is the given region. The transformation,

$$x = u^4, \qquad y = v^4, \qquad z = w^4,$$

converts S to that part of the sphere $u^2 + v^2 + w^2 \leqslant b^2$ (writing $a = b^4$)

which lies in the first octant. The Jacobian of the transformation is:

$$\frac{\partial(x, y, z)}{\partial(u, v, w)} = \begin{vmatrix} 4u^3 & 0 & 0 \\ 0 & 4v^3 & 0 \\ 0 & 0 & 4w^3 \end{vmatrix}$$

$$= 4^3 u^3 v^3 u^3.$$

Hence,

$$\iiint_S dx\, dy\, dz = 4^3 \iiint u^3 v^3 w^3\, du\, dv\, dw.$$

Convert to spherical coordinates:

$$u = \rho \sin \phi \cos \theta,\quad v = \rho \sin \phi \sin \theta,\quad w = \rho \cos \phi,$$

$$\frac{\partial(u, v, w)}{\partial(\rho, \phi, \theta)} = \rho^2 \sin \phi.$$

Then the right-hand integral is:

$$4^3 \int_0^{\pi/2} \int_0^{\pi/2} \int_0^b \rho^3 \sin^3 \phi \cos^3 \theta \,.\, \rho^3 \sin^3 \phi \sin^3 \theta \,.\, \rho^3 \cos^3 \phi \times$$

$$\rho^2 \sin \phi\, d\rho\, d\phi\, d\theta$$

$$= 4^3 \int_0^b \rho^{11}\, d\rho \,.\, \int_0^{\pi/2} \sin^7 \phi \cos^3 \phi\, d\phi \,.\, \int_0^{\pi/2} \sin^3 \theta \cos^3 \theta\, d\theta$$

$$= 4^3 \,.\, \frac{1}{12} b^{12} \quad \frac{2}{10}\ \frac{1}{8}\ \frac{2}{6}\ \frac{1}{4}$$

$$= \frac{a^3}{90}.$$

EXERCISES 5

(1) For each of the following transformations, compute the Jacobian and determine and describe the region onto which the circle $x^2 + y^2 \leqslant 1$ is transformed.

(i) $x - u + a, \quad y = v + b$;

(ii) $x = au, \quad y = vb$;

(iii) $x = u \cos \alpha - v \sin \alpha, \quad y = u \sin \alpha + v \cos \alpha$;

(iv) $x = au + bv, \quad y = cu + dv;$

(v) $x = \dfrac{u}{u^2 + v^2}, \quad y = \dfrac{v}{u^2 + v^2}.$

(2) Evaluate:

(i) $\iint x \, dA,$

(ii) $\iint x^2 y^2 \, dA,$

(iii) $\iint (x^2 + y^2) \, dA,$

where in each case the region of integration is that part of the interior of the ellipse $(x^2/a^2) + (y^2/b^2) = 1$ lying in the first quadrant.

(3) Evaluate

(i) $\iiint xyz \, dV,$

(ii) $\iiint e^{\sqrt{(x^2/a^2 + y^2/b^2 + z^2/c^2)}} \, dV,$

where in each case the region of integration is that part of the interior of the ellipsoid $(x^2/a^2) + (y^2/b^2) + (z^2/c^2) = 1$ lying in the first octant.

(4) Show that

$$\iint_R (x^2 - y^2)^2 \, dA = \frac{4}{45},$$

where R is the triangle bounded by the lines $y = 0, x = 1, x = y,$

(i) by computing directly a corresponding repeated integral,

(ii) by introducing the transformation $u = x + y, v = x - y.$

(5) Use the substitution:

$$u = y^2/x, \qquad v = xy^2,$$

to show that the area in the first quadrant bounded by the curves $y^2 = 4x$, $y^2 = 9x, xy^2 = 1, xy^2 = 4$, is $(24 + 4\sqrt{3} - 6\sqrt{2} - 8\sqrt{6})/9$.

(6) Find the volume of the solid bounded by the six parabolic cylinders $x^2 = ay, x^2 = 2ay, y^2 = az, y^2 = 2az, z^2 = ax, z^2 = 2ax$.

(7) Evaluate:

$$\iint_R x \, dA,$$

where R is the region bounded by the curves $x(1-y) = 1, x(1-y) = 2$, $xy = 1$, and $xy = 3$.

(8) For the mapping

$$u = \frac{y}{x^3}, \quad v = \frac{x}{y^3},$$

show that

$$\frac{\partial(u, v)}{\partial(x, y)} = (4uv)^{3/2}$$

Find the area in the first quadrant bounded by the curves $y = ax^3, y = bx^3$, $x = cy^3, x = dy^3$ $(b > a > 0, d > c > 0)$.

(9) Evaluate:

$$\iint e^{(y-x)/(y+x)} \, dA$$

over the triangle bounded by $x = 0, y = 0, x + y = 1$.

(10) Evaluate the repeated integral:

$$\int_1^2 \int_x^{4-x} x(x+y) \, e^{x^2} \, dy \, dx$$

and verify your result by showing that, under the transformation

$$u = x^2, \qquad v = x + y,$$

the integral becomes

$$\tfrac{1}{2} \int_1^4 \int_{2\sqrt{u}}^4 v e^u \, dv \, du$$

and then evaluating the integral in this form.

(11) Integrate the function $1/xy$ over the region bounded by the four circles $x^2 + y^2 = ax$, $x^2 + y^2 = a'x$, $x^2 + y^2 = by$, $x^2 + y^2 = b'y$, where a, a', b, b' are positive.

(12) Use the substitution:

$$x = r\cos^2\theta, \qquad y = r\sin^2\theta,$$

to evaluate:

$$\iint \frac{x^2 - y^2}{\sqrt{(xy)}}\,\mathrm{d}A$$

over the trapezium bounded by the lines $x + y = a$, $x + y = b$, $y = x\tan^2\alpha$, $y = x\tan^2\beta$, where $0 < a < b$, $0 < \alpha < \beta < \pi/2$.

(13) Evaluate:

$$\iint (4x^2 + 2xy + 3y^2 + 2x - 5y + 4)\,\mathrm{d}A$$

over the interior of the circle $x^2 + y^2 + 4x - 6y - 12 = 0$.

(14) Show that, by means of the transformation,

$$x = u(1 + v), \qquad y = v(1 + u),$$

the integral,

$$\int_0^2 \int_0^x \left\{(x - y)^2 + 2(x + y) + 1\right\}^{-\frac{1}{2}}\,\mathrm{d}y\,\mathrm{d}x,$$

becomes

$$\int_0^1 \int_v^{2/(1+v)} \mathrm{d}u\,\mathrm{d}v,$$

and hence evaluate the former integral.

6
Further Topics

Integral estimates

We list here some basic results for multiple integrals which have well-known analogues for ordinary definite integrals. For simplicity, the results will be stated only for double integrals.

In general in this book we have been assuming that $f(x, y)$ is a continuous function of x and y and that the region R is closed (that is, the boundary is included as part of R). These conditions are sufficient to ensure that the limit by which a double integral is defined exists and is finite. They are in fact more than sufficient: double integrals exist under more general conditions. But it is not our aim to go into such questions in any detail. These remarks are made at this time only to emphasise the necessity of the conditions in (v) and (vi) below.

(i) If $f(x, y) \geqslant 0$ throughout R, then

$$\iint_R f(x, y)\, dA \geqslant 0,$$

and if $f(x, y) \leqslant 0$ throughout R,

$$\iint_R f(x, y)\, dA \leqslant 0.$$

This follows from the definition of the double integral.

(ii) If $f(x, y) \leqslant g(x, y)$ throughout R, then $f(x, y) - g(x, y) \leqslant 0$ and (i) implies that

$$\iint_R f(x, y)\, dA \leqslant \iint_R g(x, y)\, dA.$$

(iii) In particular, if m and M are constants such that $m \leqslant f(x, y) \leqslant M$ throughout R, then we obtain

$$mA \leqslant \iint_R f(x, y)\, dA \leqslant MA,$$

where A is the area of R.

(iv) If, as usual, $|f(x, y)|$ denotes the absolute value of $f(x, y)$, then certainly $-|f(x, y)| \leqslant f(x, y) \leqslant |f(x, y)|$, and (ii) implies that

$$-\iint_R |f(x, y)|\, dA \leqslant \iint_R f(x, y)\, dA \leqslant \iint_R |f(x, y)|\, dA.$$

This important result is better stated in the form:

$$\left|\iint_R f(x, y)\, dA\right| \leqslant \iint_R |f(x, y)|\, dA.$$

(v) Now suppose that $f(x, y)$ is a continuous function throughout R and that R is closed. Then it can be shown that $f(x, y)$ takes on its least and greatest values at least once in R. If these values are given to m and M in (iii), then the inequalities there imply the existence of some number μ such that

$$\iint_R f(x, y)\, dA = \mu A, \qquad m \leqslant \mu \leqslant M.$$

Furthermore, the value μ will in fact be taken on by $f(x, y)$ at at least one point (X, Y) in R so that we may write

$$\iint_R f(x, y)\, dA = f(X, Y) \,.\, A.$$

In general the points (X, Y) cannot be precisely determined. This result is called the mean-value theorem for multiple integrals.

(vi) It is not difficult to prove the following extension of the mean-value theorem. If $f(x, y)$ is a continuous function in R and $p(x, y)$ is a continuous function in R which is always positive, then at least one point (X, Y) in R exists such that

$$\iint_R p(x, y)\, f(x, y)\, dA = f(X, Y) \iint_R p(x, y)\, dA.$$

Vanishing Jacobians

In the preceding chapter concerning change of variables, we mentioned that the transformation must be such that neither its Jacobian nor that of its inverse becomes zero anywhere in the region of integration. Since our regions are always closed, this implies that the Jacobians cannot become zero even on the boundary of the region.

The point was not stressed in Chapter 5 (indeed, in a number of examples and exercises we broke this rule), because in practice this does not often provide any inconvenience and in such cases can be successfully ignored (as we did). However, we will consider the matter a little more deeply now since the methods we will use suggest the approach to be adopted in the next section on improper integrals, in which the corresponding difficulties cannot always be ignored.

We obtain sufficient insight into the matter by considering the integral

$$\iint_R xy \, \mathrm{d}A,$$

where R is the quarter-circle defined by $x^2 + y^2 \leqslant 1, x \geqslant 0, y \geqslant 0$. The integral will be evaluated in three different ways.

(i) The region R is shown in Fig. 28(*a*). It may be expressed by the inequalities $0 \leqslant x \leqslant 1, 0 \leqslant y \leqslant \sqrt{(1-x^2)}$. So,

$$\begin{aligned}\iint_R xy \, \mathrm{d}A &= \int_0^1 \int_0^{\sqrt{(1-x^2)}} xy \, \mathrm{d}y \, \mathrm{d}x \\ &= \int_0^1 x \left[\tfrac{1}{2} y^2\right]_0^{\sqrt{(1-x^2)}} \mathrm{d}x \\ &= \tfrac{1}{2} \int_0^1 x(1-x^2) \, \mathrm{d}x \\ &= \tfrac{1}{8}.\end{aligned}$$

(ii) Alternatively, we might like to use polar coordinates. Then R is described by the inequalities $0 \leqslant \theta \leqslant \pi/2, 0 \leqslant r \leqslant 1$. The Jacobian of the transformation $x = r \cos \theta, y = r \sin \theta$, is r, and this vanishes when $r = 0$, at the origin. We remove this difficulty by enclosing the origin in a small circle of radius ϵ. Call its intersection with R C_ϵ. The rest of R may be denoted by $R - C_\epsilon$ (Fig. 28(*b*).)

In C_ϵ, $0 \leqslant x \leqslant \epsilon$ and $0 \leqslant y \leqslant \epsilon$, so $0 \leqslant xy \leqslant \epsilon^2$. The area of C_ϵ is $\frac{1}{4} \pi\epsilon^2$.

Using the estimate in (iii) of the preceding section,

$$0 \leqslant \iint_{C_\epsilon} xy \, dA \leqslant \tfrac{1}{4}\pi\epsilon^4 .$$

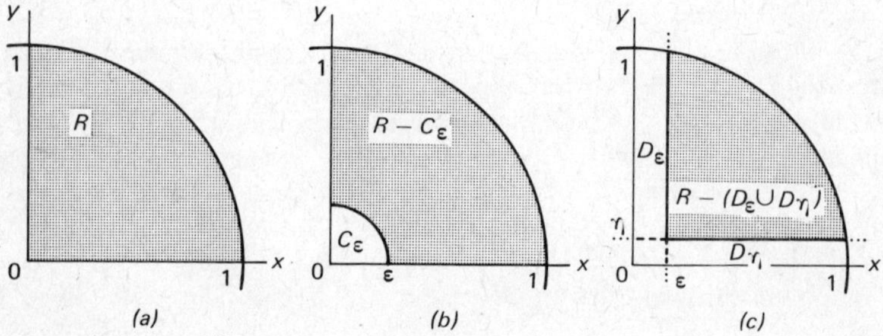

Fig. 28

Thus $\iint_{C_\epsilon} xy \, dA \to 0$ as $\epsilon \to 0$.

$R - C_\epsilon$ is described by the inequalities $0 \leqslant \theta \leqslant \pi/2$, $\epsilon \leqslant r \leqslant 1$. Since

$$\iint_R xy \, dA = \iint_{R - C_\epsilon} xy \, dA + \iint_{C_\epsilon} xy \, dA,$$

we thus have:

$$\iint_R xy \, dA = \lim_{\epsilon \to 0} \iint_{R - C_\epsilon} xy \, dA$$

$$= \lim_{\epsilon \to 0} \int_0^{\pi/2} \int_\epsilon^1 r \cos\theta \, . \, r \sin\theta \, . \, r \, dr \, d\theta$$

$$= \int_0^{\pi/2} \cos\theta \sin\theta \, d\theta \, . \, \lim_{\epsilon \to 0} \int_\epsilon^1 r^3 \, dr$$

$$= \frac{1}{2} \lim_{\epsilon \to 0} \frac{1}{4}(1 - \epsilon^4).$$

$$= \frac{1}{8}.$$

(iii) A third method is to use the transformation:

$$x = \sqrt{u}, \qquad y = \sqrt{v}.$$

Then R is transformed into the triangle bounded by $u = 0$, $v = 0$ and $u + v = 1$. The Jacobian is:

$$\frac{\partial(x, y)}{\partial(u, v)} = \begin{vmatrix} \frac{1}{2\sqrt{u}} & 0 \\ 0 & \frac{1}{2\sqrt{v}} \end{vmatrix}$$

$$= \frac{1}{4\sqrt{(uv)}}.$$

The Jacobian of the inverse transformation $u = x^2$, $v = y^2$ is thus $4\sqrt{(uv)} = 4xy$ and this vanishes whenever $x = 0$ or $y = 0$: along all of the two straight boundaries of R.

We must remove these lines from R. This is done by introducing the almost-rectangles D_ϵ and D_η shown in Fig. 28(c). The rest of R may be denoted by $R - (D_\epsilon \cup D_\eta)$. In D_ϵ, $0 \leqslant x \leqslant \epsilon$ and $0 \leqslant y \leqslant 1$, so $0 \leqslant xy \leqslant \epsilon$. The area of D_ϵ is less than ϵ, so we have the estimate:

$$0 \leqslant \iint_{D_\epsilon} xy \, \mathrm{d}A \leqslant \epsilon^2.$$

Thus $\iint_{D_\epsilon} xy \, \mathrm{d}A \to 0$ as $\epsilon \to 0$. In a similar way, $\iint_{D_\eta} xy \, \mathrm{d}A \to 0$ as $\eta \to 0$.

Certainly also $\iint_{D_\epsilon \cap D_\eta} xy \, \mathrm{d}A \to 0$ as both $\epsilon \to 0$ and $\eta \to 0$. Writing

$$\iint_R xy \, \mathrm{d}A = \iint_{R - (D_\epsilon \cup D_\eta)} xy \, \mathrm{d}A + \iint_{D_\epsilon} xy \, \mathrm{d}A +$$

$$+ \iint_{D_\eta} xy \, \mathrm{d}A - \iint_{D_\epsilon \cap D_\eta} xy \, \mathrm{d}A$$

we thus see that

$$\iint_R xy \, \mathrm{d}A = \lim_{\substack{\epsilon \to 0 \\ \eta \to 0}} \iint_{R - (D_\epsilon \cup D_\eta)} xy \, \mathrm{d}A.$$

$R - (D_\epsilon \cup D_\eta)$ becomes, under the given transformation, the triangle defined by $\epsilon^2 \leqslant u \leqslant 1 - \eta^2$, $\eta^2 \leqslant v \leqslant 1 - u$ (see Fig. 29), and

$$\iint_{R-(D_\epsilon \cup D_\eta)} xy \, \mathrm{d}A = \int_{\epsilon^2}^{1-\eta^2} \int_{\eta^2}^{1-u} \sqrt{(uv)} \cdot \frac{1}{4\sqrt{(uv)}} \, \mathrm{d}v \, \mathrm{d}u$$

$$= \tfrac{1}{4} \int_{\epsilon^2}^{1-\eta^2} (1 - u - \eta^2) \, \mathrm{d}u$$

$$= \tfrac{1}{4} \cdot \tfrac{1}{2} \, (1 - \epsilon^2 - \eta^2)^2$$

$$\to \tfrac{1}{8},$$

as $\epsilon \to 0$ and $\eta \to 0$.

These examples illustrate the method of 'removing' awkward points or lines (or curves) from a region. A corresponding discussion could be given for triple integrals. The situation is necessarily more complex: it may be required to remove points, lines, curves, planes or surfaces from the solid region of integration.

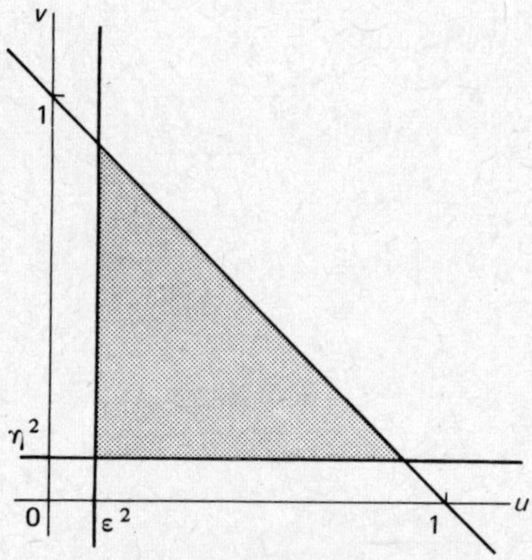

Fig. 29

There is one other aspect of the vanishing Jacobian that is worth pointing out. Suppose we are to find the integral,

$$\iint_R \left(\frac{x^2 + y^2}{xy}\right)^2 \mathrm{d}A,$$

where R is the region common to the circles $x^2 + y^2 \leqslant ax$ and $x^2 + y^2 \leqslant by$ (a, b positive). Let us use the transformation

$$u = \frac{x^2 + y^2}{x}, \qquad v = \frac{x^2 + y^2}{y}.$$

Then,

$$\frac{\partial(u, v)}{\partial(x, y)} = \begin{vmatrix} \dfrac{x^2 - y^2}{x^2} & \dfrac{2y}{x} \\ \\ \dfrac{2x}{y} & \dfrac{y^2 - x^2}{y^2} \end{vmatrix}$$

$$= -\left(\frac{x^2 + y^2}{xy}\right)^2,$$

and we get the neat result:

$$\iint_R \left(\frac{x^2 + y^2}{xy}\right)^2 \mathrm{d}x\, \mathrm{d}y = \iint_{R'} \mathrm{d}u\, \mathrm{d}v.$$

R' is the region in the uv-plane to which R is transformed. The right-hand side of the above equation shows that we require the area of R'. But here is the problem. The circle $x^2 + y^2 = ax$ is transformed into the line $u = a$ and the circle $x^2 + y^2 = by$ is transformed into the line $v = b$. Where is R'?

Again the difficulty lies in the value of the Jacobian at the origin. We have

$$\frac{\partial(x, y)}{\partial(u, v)} = -\left(\frac{xy}{x^2 + y^2}\right)^2,$$

and

$$\lim_{x \to 0}\left(\lim_{y \to 0} \frac{xy}{x^2 + y^2}\right) = 0 \quad \text{and} \quad \lim_{y \to 0}\left(\lim_{x \to 0} \frac{xy}{x^2 + y^2}\right) = 0.$$

These imply that the limiting value of the Jacobian $\partial(x, y)\, \partial(u, v)$ as we approach the origin along either the x- or y-axis is zero. But if we approach the origin along the line $y = x$ we obtain a limiting value of $-\frac{1}{4}$. The situation is therefore somewhat worse than the case of a vanishing Jacobian. We can in fact give no value to the Jacobian at the origin. But the effect is the same: we must exclude the origin from R.

The best way to do this is via the circles $x^2 + y^2 = \epsilon x$ and $x^2 + y^2 = \eta y$ where ϵ and η are small and positive. We then integrate over the shaded region shown in Fig. 30. These circles are transformed into the lines $u = \epsilon$ and $v = \eta$, respectively, in the uv-plane. We thus require

$$\iint \mathrm{d}u\, \mathrm{d}v$$

over the rectangle bounded by the lines $u = \epsilon$, $u = a$, $v = \eta$, $v = b$. This has area $(a - \epsilon)(b - \eta)$.

Hence

$$\iint_R \left(\frac{x^2+y^2}{xy}\right)^2 \mathrm{d}A = \lim_{\substack{\epsilon \to 0 \\ \eta \to 0}} (a-\epsilon)(b-\eta)$$

$$= ab.$$

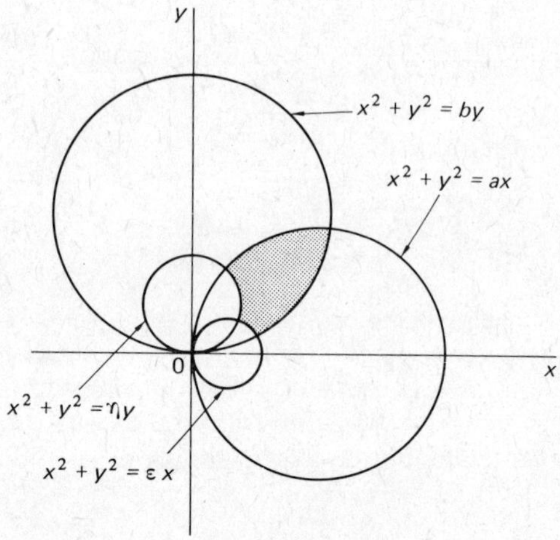

Fig. 30

Improper integrals

The situation is well illustrated by considering the integral

$$I \equiv \iint_R \frac{x-y}{(x+y)^3} \, \mathrm{d}A,$$

where R is the square $0 \leqslant x \leqslant 1$, $0 \leqslant y \leqslant 1$.

In the usual way, we convert the double integral into the repeated integral

$$\int_0^1 \int_0^1 \frac{x-y}{(x+y)^3} \, \mathrm{d}y \, \mathrm{d}x.$$

The inner integral may be computed using integration by parts:

$$\int_0^1 \frac{x-y}{(x+y)^3} \, \mathrm{d}y = \int_0^1 (y-x) \, \mathrm{d}\left\{\tfrac{1}{2}(x+y)^{-2}\right\}$$

$$= \left[\tfrac{1}{2}(x+y)^{-2}(y-x) \right]_{y=0}^{y=1} - \tfrac{1}{2}\int_0^1 (x+y)^{-2}\, dy$$

$$= \frac{1-x}{2(1+x)^2} + \frac{1}{2x} + \frac{1}{2}\left[(x+y)^{-1} \right]_{y=0}^{y=1}$$

$$= \frac{1-x}{2(1+x)^2} + \frac{1}{2x} + \frac{1}{2(1+x)} - \frac{1}{2x}$$

$$= \frac{1}{(1+x)^2}.$$

Then,

$$I = \int_0^1 \frac{1}{(1+x)^2}\, dx$$

$$= \left[-(1+x)^{-1} \right]_0^1$$

$$= \tfrac{1}{2}.$$

However, suppose the integrations were performed in the opposite order; that is, consider the repeated integral:

$$\int_0^1 \int_0^1 \frac{x-y}{(x+y)^3}\, dx\, dy.$$

We have,

$$\int_0^1 \frac{x-y}{(x+y)^3}\, dx = \int_0^1 (y-x)\, d\left\{ \tfrac{1}{2}(x+y)^{-2} \right\}$$

$$= \left[\tfrac{1}{2}(x+y)^{-2}(y-x) \right]_{x=0}^{x=1} + \tfrac{1}{2}\int_0^1 (x+y)^{-2}\, dx$$

$$= \frac{y-1}{2(y+1)^2} - \frac{1}{2y} - \frac{1}{2(y+1)} + \frac{1}{2y}$$

$$= -\frac{1}{(y+1)^2},$$

so that

$$I = \int_0^1 \frac{-1}{(y+1)^2}\, dy$$

$$= -\tfrac{1}{2}.$$

The two procedures give different values for I! We must conclude that in fact I does not exist as a double integral. This can also be seen in the following way.

The difficulty must be at the origin, where, since both x and y are zero, the integrand is not defined. We shall therefore exclude the origin from our square, in the manner of the preceding section, and see what happens as we allow the new region to tend back towards the full square.

In Fig. 31, the origin has been excluded by constructing the rectangle formed by the lines $x = 0, y = 0, x = \epsilon, y = \eta, \epsilon > 0, \eta > 0$. The integral

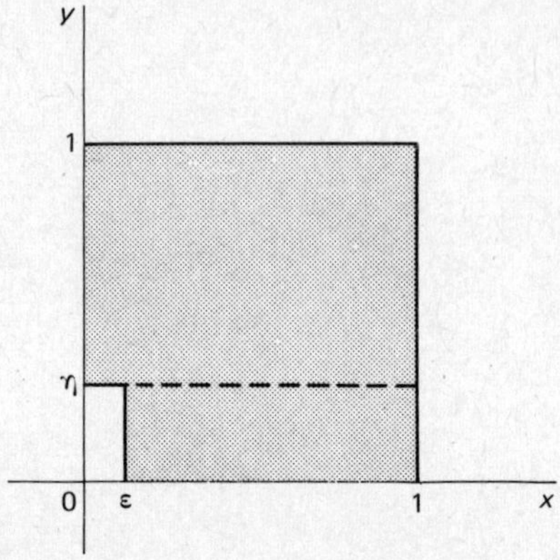

Fig. 31

over the shaded portion is evaluated in a similar way to the preceding integrations. (Note that the region must be split into two, as indicated.) We obtain:

$$\int_0^1 \int_\eta^1 \frac{x-y}{(x+y)^3}\,dy\,dx + \int_\epsilon^1 \int_0^\eta \frac{x-y}{(x+y)^3}\,dy\,dx$$

$$= \tfrac{1}{2}\int_0^1 \Big[(x+y)^{-2}(y-x) + (x+y)^{-1}\Big]_{y=\eta}^{y=1}\,dx$$

$$+ \tfrac{1}{2}\int_\epsilon^1 \Big[(x+y)^{-2}(y-x) + (x+y)^{-1}\Big]_{y=0}^{y=\eta}\,dx$$

$$= \tfrac{1}{2}\int_0^1 \left\{\frac{1-x}{(1+x)^2} + \frac{1}{1+x} - \frac{\eta-x}{(\eta+x)^2} - \frac{1}{\eta+x}\right\}dx$$

$$+ \tfrac{1}{2}\int_\epsilon^1 \left\{\frac{\eta-x}{(\eta+x)^2} + \frac{1}{\eta+x} + \frac{1}{x} - \frac{1}{x}\right\}dx$$

$$= \int_0^1 \left\{\frac{1}{(1+x)^2} - \frac{\eta}{(\eta+x)^2}\right\} dx + \int_\epsilon^1 \frac{\eta}{(\eta+x)^2} dx$$

$$= \left[\frac{-1}{1+x} + \frac{\eta}{\eta+x}\right]_0^1 + \left[\frac{-\eta}{\eta+x}\right]_\epsilon^1$$

$$= \left(-\frac{1}{2} + \frac{\eta}{\eta+1} + 1 - 1\right) + \left(\frac{-\eta}{\eta+1} + \frac{\eta}{\eta+\epsilon}\right)$$

$$= -\frac{1}{2} + \frac{\eta}{\eta+\epsilon}.$$

The integral over R is arrived at by letting ϵ and η tend to zero. But

$$\lim_{\substack{\epsilon \to 0 \\ \eta \to 0}} \frac{\eta}{\eta+\epsilon}$$

is indeterminate: since

$$\lim_{\eta \to 0} \left(\lim_{\epsilon \to 0} \frac{\eta}{\eta+\epsilon}\right) = 1,$$

while

$$\lim_{\epsilon \to 0} \left(\lim_{\eta \to 0} \frac{\eta}{\eta+\epsilon}\right) = 0,$$

no definite value can be given to the limit and hence the double integral does not exist over R.

This example also shows that we needed to exclude the origin by a rectangle, say, and not just a square. If we put $\eta = \epsilon$ in the above, we could say

$$I = \lim_{\epsilon \to 0} \left(-\frac{1}{2} + \frac{\epsilon}{2\epsilon}\right) = -\frac{1}{2} + \frac{1}{2} = 0,$$

different again from the two values obtained previously!

We shall now consider such integrals more generally, confining our remarks to double integrals. There are three types of improper integrals which we need to discuss.

(i) *Integrands with a finite number of isolated infinite discontinuities*

The example worked through above is of this type. In it there was just one point, the origin, at which the integrand became infinite. All improper integrals of this type may be considered to be of this form; for since there is only a finite number of points of infinite discontinuity, the region of integration may be split up so that within each of the resulting regions there is at most one of these points, and then, considering the integral over each region separately, by translation of axes, the point may be taken at the origin.

An analogous situation for ordinary definite integrals concerns the integral

$$\int_a^b f(x)\,dx$$

where $f(x) \to \infty$ as $x \to a$. This integral is defined as

$$\lim_{\epsilon \to 0} \int_{a+\epsilon}^{b} f(x)\,dx,$$

if this limit exists.

In our situation we now suppose that the integrand $f(x, y)$ becomes infinite at the origin and that the region R includes the origin. To maintain the analogy, we adopt the following symbolic definition:

$$\iint_R f(x, y)\,dA = \lim_{\delta \to 0} \iint_{R-\delta} f(x, y)\,dA,$$

if this limit exists. It must be understood here that δ is a domain which serves to exclude the origin from R and whose area tends to zero in such a way that its linear dimensions also tend to zero; by '$\delta \to 0$' we mean in fact the parameter or parameters by which δ is defined tend to zero; furthermore, the limit must be independent of the choice of δ.

When the limit exists, the integral is said to converge; otherwise it is said to diverge.

If the function $f(x, y)$ is positive throughout R, then $\iint_{R-\delta} f(x, y)\,dA$ must get larger as δ gets smaller. In that case, provided a constant K can be found such that $\iint_{R-\delta} f(x, y)\,dA \leqslant K$ for all δ, the convergence of the integral is assured. Any choice of δ (as long as $\delta \to 0$) will then be sufficient.

Example. Show that:

$$\iint_R \frac{dA}{(x^2 + y^2)^\alpha}$$

converges if and only if $\alpha < 1$, where R is any (finite) region containing the origin.

Solution: The integrand is always positive and becomes infinite at the origin. It is sufficient to consider convergence in the region R_0: $x^2 + y^2 \leqslant 1$ since there is no difficulty away from the origin. Any δ will do: most convenient is the circle $x^2 + y^2 \leqslant \epsilon^2 < 1$. $R_0 - \delta$ is then the annulus $\epsilon^2 \leqslant x^2 + y^2 \leqslant 1$. Converting to polar coordinates, this is described by $0 \leqslant \theta \leqslant 2\pi$, $\epsilon \leqslant r \leqslant 1$, so

$$\iint_{R_0} \frac{\mathrm{d}A}{(x^2+y^2)^\alpha} = \lim_{\delta\to 0} \iint_{R_0-\delta} \frac{\mathrm{d}A}{(x^2+y^2)^\alpha}$$

$$= \lim_{\epsilon\to 0} \int_0^{2\pi} \int_\epsilon^1 \frac{r\,\mathrm{d}r\,\mathrm{d}\theta}{r^{2\alpha}}$$

$$= \lim_{\epsilon\to 0} 2\pi \int_\epsilon^1 r^{1-2\alpha}\,\mathrm{d}r$$

$$= \frac{\pi}{1-\alpha} \lim_{\epsilon\to 0} (1 - \epsilon^{2-2\alpha}) \qquad (\alpha \neq 1)$$

This last limit exists if and only if $\alpha < 1$, and then

$$\iint_{R_0} (x^2+y^2)^{-\alpha}\,\mathrm{d}A = \pi/(1-\alpha).$$

Thus the original integral converges if and only if $\alpha < 1$. (The case $\alpha = 1$ requires a separate treatment. But then,

$$\iint_{R_0} \frac{\mathrm{d}A}{x^2+y^2} = \lim_{\epsilon\to 0} \int_0^{2\pi} \int_\epsilon^1 \frac{r\,\mathrm{d}r\,\mathrm{d}\theta}{r^2}$$

$$= -2\pi \lim_{\epsilon\to 0} \log \epsilon,$$

which is infinite.)

To treat integrals whose integrands are not always positive, we introduce the notion of absolute convergence: $\iint_R \mathrm{f}(x, y)\,\mathrm{d}A$ is said to be absolutely convergent if $\iint_R |\mathrm{f}(x, y)|\,\mathrm{d}A$ is convergent.

If an integral is absolutely convergent, then it must be convergent. To prove this, the inequalities $-|\mathrm{f}(x, y)| \leqslant \mathrm{f}(x, y) \leqslant |\mathrm{f}(x, y)|$ imply both

$$0 \leqslant \tfrac{1}{2}\{|\mathrm{f}(x, y)| + \mathrm{f}(x, y)\} \leqslant |\mathrm{f}(x, y)|,$$

and

$$0 \leqslant \tfrac{1}{2}\{|f(x, y)| - f(x, y)\} \leqslant |f(x, y)|,$$

and these in turn imply that the integrals

$$\iint_R \tfrac{1}{2}\{|f(x, y)| + f(x, y)\}\, dA$$

and

$$\iint_R \tfrac{1}{2}\{|f(x, y)| - f(x, y)\}\, dA$$

converge since the integrands are positive and

$$\iint_R |f(x, y)|\, dA$$

has been assumed to be finite. But then the difference

$$\iint_R \tfrac{1}{2}\Big\{|f(x, y)| + f(x, y)\Big\}\, dA - \iint_R \tfrac{1}{2}\Big\{|f(x, y)| - f(x, y)\Big\}\, dA$$

$$= \iint_R \Big[\tfrac{1}{2}\Big\{|f(x, y)| + f(x, y)\Big\} - \tfrac{1}{2}\Big\{|f(x, y)| - f(x, y)\Big\}\Big] dA$$

$$= \iint_R f(x, y)\, dA,$$

must converge, and this is the required result.

We can now make use of the preceding example to give a result which is useful in testing the convergence of certain double integrals:

If $g(x, y)$ is a continuous function in R which is such that $|g(x, y)| \leqslant K$ in R, then

$$\iint_R \frac{g(x, y)}{(x^2 + y^2)^\alpha}\, dA$$

converges, provided $\alpha < 1$.

This follows since

$$\iint_R (x^2 + y^2)^{-\alpha}\, dA$$

has been shown to converge, when $\alpha < 1$, and

$$\left|\frac{g(x, y)}{(x^2 + y^2)^\alpha}\right| \leqslant K \cdot \frac{1}{(x^2 + y^2)^\alpha}.$$

Thus the given integral is absolutely convergent and hence convergent.

Example. Prove that:

$$\iint_R \frac{2x^3 e^y}{\sqrt{(x^2 + y^2)}} \, dA$$

converges, whatever the (finite) region R.

Solution: If the origin is not included in R, there is no difficulty. If it is, we make use of the test just given. Since R is a finite closed region, the function $2x^3 e^y$ will take on its maximum and minimum values at certain points in it; let the larger of these, in absolute value, be K, so that $|2x^3 e^y| \leqslant K$ throughout R. With this K, and $\alpha = \frac{1}{2}$, the above test implies the convergence of the integral.

(ii) *Integrands with lines of infinite discontinuity*

It may well happen that the integrand becomes infinite not just at isolated points of the region of integration but along whole curves in it. Nonetheless, the situation may be handled almost entirely in a similar fashion to the preceding one.

As a simple illustration, consider the integral

$$\iint_R \frac{dA}{x^\alpha}, \qquad \alpha > 0,$$

where R is the square $0 \leqslant x \leqslant 1, 0 \leqslant y \leqslant 1$. The integrand becomes infinite when $x = 0$: along the entire left-hand edge of the square. This edge may be removed from R by the line $x = \epsilon, 0 < \epsilon < 1$, as in Fig. 32. We integrate over the remaining (shaded) portion:

$$\int_0^1 \int_\epsilon^1 \frac{1}{x^\alpha} \, dx \, dy$$

$$= \frac{1}{1 - \alpha} \left[x^{1-\alpha} \right]_\epsilon^1 \qquad (\alpha \neq 1)$$

$$= \frac{1}{1 - \alpha} (1 - \epsilon^{1-\alpha}).$$

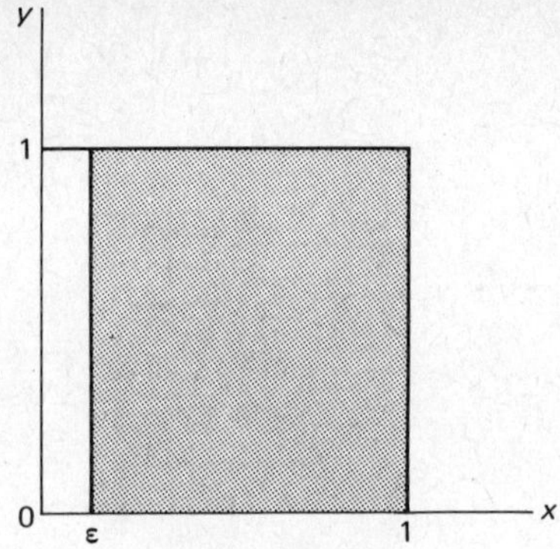

Fig. 32

As ϵ tends to zero, this will approach a finite limit provided $\alpha < 1$. Thus,

$$\iint_R \frac{dA}{x^\alpha} = \frac{1}{1-\alpha} \lim_{\epsilon \to 0} (1 - \epsilon^{1-\alpha})$$

$$= \frac{1}{1-\alpha},$$

if $\alpha < 1$. (Since the integrand is non-negative in R, the same limit would be approached no matter how the edge was excluded.) In this case, we again adopt the definition,

$$\iint_R f(x, y)\, dA = \lim_{\delta \to 0} \iint_{R-\delta} f(x, y)\, dA,$$

if the limit exists. Here δ is a domain which serves to exclude the offending curve of discontinuity from R and whose area tends to zero; the limit must be independent of the choice of δ.

(iii) *Infinite regions of integration*

To date, all the regions of integration which we have considered have been finite. We now consider multiple integrals which are improper in the sense that their regions of integration extend to infinity in some way. For ordinary definite integrals, these correspond to integrals such as

$$\int_{-\infty}^{\infty} f(x)\, dx$$

and

$$\int_a^{\infty} f(x)\, dx.$$

The latter integral is defined as

$$\lim_{M \to \infty} \int_a^M f(x)\, dx,$$

if this limit exists; we define corresponding double integrals in an analogous fashion.

For any properly defined region R_M (for example, the circle $x^2 + y^2 \leqslant M^2$ or the square $-M \leqslant x \leqslant M$, $-M \leqslant y \leqslant M$) the region $R_M \cap R$ is a finite region which, as M increases, may be thought of as giving increasingly better approximations to the infinite region R. Accordingly, we define

$$\iint_R f(x, y)\, dA$$

by

$$\iint_R f(x, y)\, dA = \lim_{M \to \infty} \iint_{R_M \cap R} f(x, y)\, dA,$$

if this limit exists (in which case again we call the integral convergent). The regions R_M may be defined explicitly in terms of a parameter M (as in the above examples) or they may be defined in terms of more than one parameter: by '$M \to \infty$' we mean all relevant parameters tend to infinity; furthermore the limit must be independent of the choice of the approximating regions.

For integrals in which the integrand is always positive, it is easy to show that any suitable choice of R_M will do. In the following example, we illustrate the idea of the region R_M and evaluate a definite integral which cannot be evaluated by elementary methods.

Example. Show that:

$$\iint_R e^{-(x^2 + y^2)}\, dA = \pi/8,$$

where R is the infinite sector $0 \leqslant \theta \leqslant \pi/4$, and hence evaluate

$$\int_0^\infty e^{-x^2}\, dx.$$

Solution: We shall choose for the region R_M the square $0 \leqslant x \leqslant M$, $0 \leqslant y \leqslant M$. R is the region shown lightly shaded in Fig. 33, so $R_M \cap R$ is the triangle shown cross-hatched. It will be seen that this triangle 'approximates' R better and better as M increases.

In polar coordinates, the line $x = M$ has equation $r = M \sec\theta$, so

$$\iint_{R_M \cap R} e^{-(x^2+y^2)}\, dA = \int_0^{\pi/4} \int_0^{M \sec\theta} e^{-r^2} r\, dr\, d\theta$$

$$= \int_0^{\pi/4} \left[-\tfrac{1}{2} e^{-r^2} \right]_0^{M \sec\theta} d\theta$$

$$= \tfrac{1}{2} \int_0^{\pi/4} (1 - e^{-M^2 \sec^2\theta})\, d\theta$$

$$= \frac{\pi}{8} - \frac{1}{2} \int_0^{\pi/4} e^{-M^2 \sec^2\theta}\, d\theta.$$

The integral

$$\int_0^{\pi/4} e^{-M^2 \sec^2\theta}\, d\theta$$

is positive and may be easily estimated since $\sec^2\theta \geqslant 1$:

$$0 < \int_0^{\pi/4} e^{-M^2 \sec^2\theta}\, d\theta \leqslant \int_0^{\pi/4} e^{-M^2}\, d\theta = \frac{\pi}{4} e^{-M^2}.$$

Since the latter quantity tends to zero as M tends to infinity, we have:

$$\iint_R e^{-(x^2+y^2)}\, dA = \lim_{M\to\infty} \iint_{R_M \cap R} e^{-(x^2+y^2)}\, dA$$

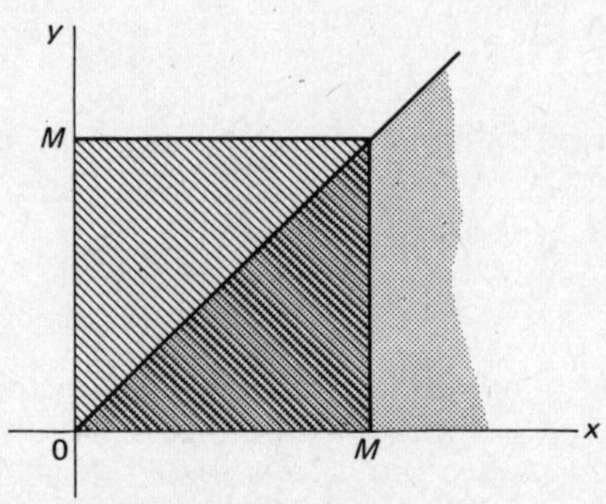

Fig. 33

$$=\frac{\pi}{8}-\frac{1}{2}\lim_{M\to\infty}\int_0^{\pi/4} e^{-M^2\sec^2\theta}\,d\theta$$

$$=\frac{\pi}{8},$$

as required.

In Cartesian coordinates,

$$\iint_{R_M \cap R} e^{-(x^2+y^2)}\,dA = \int_0^M \int_0^x e^{-(x^2+y^2)}\,dy\,dx.$$

Interchanging the order of integration,

$$\int_0^M \int_0^x e^{-(x^2+y^2)}\,dy\,dx = \int_0^M \int_y^M e^{-(x^2+y^2)}\,dx\,dy.$$

x and y are just dummy variables, so writing x for y and y for x, we have

$$\int_0^M \int_y^M e^{-(x^2+y^2)}\,dx\,dy = \int_0^M \int_x^M e^{-(y^2+x^2)}\,dy\,dx,$$

so

$$\int_0^M \int_0^x e^{-(x^2+y^2)}\,dy\,dx = \int_0^M \int_x^M e^{-(x^2+y^2)}\,dy\,dx.$$

(We have shown that the integral over the shaded area in Fig. 33 equals the integral over the cross-hatched area. This may also be shown by a direct computation.) Hence,

$$\int_0^M \int_0^M e^{-(x^2+y^2)}\,dy\,dx$$

$$=\int_0^M \left\{\int_0^x e^{-(x^2+y^2)}\,dy + \int_x^M e^{-(x^2+y^2)}\,dy\right\}dx$$

$$=2\int_0^M \int_0^x e^{-(x^2+y^2)}\,dy\,dx.$$

On the other hand,

$$\int_0^M \int_0^M e^{-(x^2+y^2)}\, dy\, dx = \int_0^M \int_0^M e^{-x^2} e^{-y^2}\, dy\, dx$$

$$= \int_0^M e^{-x^2}\, dx \,.\, \int_0^M e^{-y^2}\, dy$$

$$= \left\{\int_0^M e^{-x^2}\, dx\right\}^2,$$

since y is only a dummy variable and may be replaced by x. Thus,

$$\int_0^M e^{-x^2}\, dx = \sqrt{\left\{2\int_0^M \int_0^x e^{-(x^2+y^2)}\, dy\, dx\right\}}.$$

Letting M tend to infinity, we arrive at the result

$$\int_0^\infty e^{-x^2}\, dx = \frac{\sqrt{\pi}}{2}.$$

EXERCISES 6a

(1) (i) Prove that if $g(x, y)$ is a continuous function in R which is such that $|g(x, y)| \leqslant K$ in R, then

$$\iint_R \frac{g(x, y)}{x^\alpha}\, dA$$

converges, provided $\alpha < 1$.

(ii) Deduce a test corresponding to that in (i) for the integral

$$\iint_R \frac{g(x, y)}{x^\alpha y^\beta}\, dA.$$

(2) Prove that the following integrals converge for any finite region R:

(i) $$\iint_R \frac{\sin x}{\sqrt[3]{(x^2+y^2)}}\, dA,$$

(ii) $$\iint_R \frac{\sin(x+y)}{\sqrt{x}}\, \mathrm{d}A,$$

(iii) $$\iint_R \frac{x^x e^y}{x^{2/3} y^{1/3}}\, \mathrm{d}A.$$

(3) Evaluate:

$$\iint_R \frac{\mathrm{d}A}{(x+y)^\alpha}$$

(i) where R is the triangle bounded by $x = 0, y = 0, x + y = 1$;

(ii) where R is the parallelogram bounded by $y = 1, x + y = 1, y = -1, x + y = 0$; stating in each case the range of values of α for which the integral converges.

(4) In a manner corresponding to that in the text for double integrals, show that the triple integral,

$$\iiint_S \frac{\mathrm{d}V}{(x^2 + y^2 + z^2)^\alpha},$$

where S is any finite solid region containing the origin, converges if and only if $\alpha < 3/2$.

(5) Evaluate:

$$\int_0^\infty \int_y^\infty \frac{e^{-x}}{x}\, \mathrm{d}x\, \mathrm{d}y.$$

(6) By means of the substitution $u = x + y$, $v = y$, prove that:

$$\int_0^\infty \int_0^\infty e^{-(x+y)} \sin\left(\frac{\pi y}{x+y}\right) \mathrm{d}x\, \mathrm{d}y = 2/\pi.$$

(7) Evaluate:

$$\iint_R e^{-(x+y)^2}\, \mathrm{d}A$$

where R is the positive quadrant of the xy-plane.

(8) Use polar coordinates to show that:

$$\iint_R e^{-(x^2+2xy\cos\alpha+y^2)}\,dA = \tfrac{1}{2}\int_0^{\pi/2} \frac{d\theta}{1+\sin 2\theta\cos\alpha},$$

$$0<\alpha<\pi,$$

where R is the positive quadrant of the xy-plane, and show that this integral equals $\alpha/(2\sin\alpha)$.

(9) By considering

$$\iint_R e^{-xy}\,dA,$$

where R is the rectangle $0 \leqslant x \leqslant M$, $0 < a \leqslant y \leqslant b$, as a repeated integral in two ways, and letting M tend to infinity, show that:

$$\int_0^\infty \frac{e^{-ax}-e^{-bx}}{x}\,dx = \log(b/a), \quad a>0, b>0.$$

The gamma and beta functions

The improper definite integral,

$$\int_0^\infty e^{-x}x^{n-1}\,dx,$$

occurs in many applications of mathematics to physics and statistics. By considering separately the integrals

$$\int_0^1 e^{-x}x^{n-1}\,dx$$

and

$$\int_1^\infty e^{-x}x^{n-1}\,dx,$$

the original integral can be shown to converge for all positive values of n. It is a function of the parameter n, called the gamma function, denoted

$\Gamma(n)$. That is,

$$\Gamma(n) = \int_0^\infty e^{-x} x^{n-1}\, dx, \quad n > 0.$$

Closely related to the gamma function, though not obviously so, is the beta function, defined as a function of two parameters m and n by:

$$B(m, n) = \int_0^1 x^{m-1}(1-x)^{n-1}\, dx, \quad m > 0,\ n > 0.$$

Among the numerous alternative forms for the gamma and beta functions, we note the following:

(i) $\Gamma(n) = 2\int_0^\infty e^{-x^2} x^{2n-1}\, dx,$

for putting $x = u^2$ in the original definition, we get:

$$\Gamma(n) = \int_0^\infty e^{-u^2} u^{2(n-1)}\, 2u\, du$$

$$= 2\int_0^\infty e^{-u^2} u^{2n-1}\, du,$$

and we then change the dummy variable from u to x.

(ii) $\Gamma(n) = \int_0^1 (-\log x)^{n-1}\, dx,$

obtained by the substitution $x = -\log u$.

(iii) $B(m, n) = \int_0^1 x^{n-1}(1-x)^{m-1}\, dx,$

easily obtained by putting $x = 1 - u$. Thus,

$$B(m, n) = B(n, m).$$

(iv) Setting $x = \dfrac{u}{1+u}$ will lead to

$$B(m, n) = \int_0^\infty \frac{x^{m-1}}{(1+x)^{m+n}}\, dx,$$

so also, by (iii),

$$B(m, n) = \int_0^\infty \frac{x^{n-1}}{(1+x)^{m+n}}\, dx.$$

Writing the former result as,

$$B(m, n) = \int_0^1 \frac{x^{m-1}}{(1+x)^{m+n}}\, dx + \int_1^\infty \frac{x^{m-1}}{(1+x)^{m+n}}\, dx$$

and putting $x = 1/u$ in the second integral on the right will lead to:

$$B(m, n) = \int_0^1 \frac{x^{m-1} + x^{n-1}}{(1+x)^{m+n}}\, dx.$$

In this form, the conclusion in (iii) is obvious.

(v) In the original formulation of $B(m, n)$, put $x = \sin^2 \theta$. We then see that,

$$B(m, n) = 2 \int_0^{\pi/2} \sin^{2m-1}\theta \cos^{2n-1}\theta \, d\theta.$$

Notice that

$$\Gamma(1) = \int_0^\infty e^{-x}\, dx$$

$$= \Big[-e^{-x}\Big]_0^\infty$$

$$= 1,$$

and that, integrating by parts,

$$\Gamma(n+1) = \int_0^\infty e^{-x} x^n \, dx$$

$$= -\int_0^\infty x^n \, d(e^{-x})$$

$$= -\Big[e^{-x} x^n\Big]_0^\infty + \int_0^\infty e^{-x} . n x^{n-1}\, dx$$

$$= 0 + n \int_0^\infty e^{-x} x^{n-1}\, dx$$

$$= n\, \Gamma(n)$$

(since $\lim_{x \to \infty} e^{-x} x^n = 0$ for any constant n), provided $n > 0$.

In the particular case when n is a positive integer, we now have

$$\begin{aligned} \Gamma(n+1) &= n\, \Gamma(n) \\ &= n\,(n-1)\,\Gamma(n-1) \\ &= \ldots \\ &= n(n-1)(n-2)\ldots 3\,.\,2\,.\,1\;\Gamma(1) \\ &= n(n-1)(n-2)\ldots 3\,.\,2\,.\,1, \end{aligned}$$

that is $\Gamma(n+1) = n!$, a remarkable result.

The recurrence formula $\Gamma(n+1) = n\,\Gamma(n)$ is of course applicable for any positive n, not just integers. From the formulation (i), we see that

$$\begin{aligned} \Gamma(\tfrac{1}{2}) &= 2 \int_0^\infty e^{-x^2}\, dx \\ &= \sqrt{\pi}, \end{aligned}$$

since we evaluated the important integral $\int_0^\infty e^{-x^2}\, dx$ in the previous section. Then, for example,

$$\Gamma\left(\frac{3}{2}\right) = \frac{1}{2}\Gamma\left(\frac{1}{2}\right) = \frac{1}{2}\sqrt{\pi},$$

$$\Gamma\left(\frac{5}{2}\right) = \frac{3}{2}\Gamma\left(\frac{3}{2}\right) = \frac{3}{4}\sqrt{\pi}.$$

The link between the gamma and beta functions is arrived at by considering the double integral,

$$I = 4 \iint_R e^{-(x^2 + y^2)} x^{2m-1} y^{2n-1}\, dA,$$

where R is the whole positive quadrant.

In discussing the convergence of this integral we may make any suitable choice for the approximating regions R_M, since the integrand is always positive. Then,

$$I = \lim_{M\to\infty} 4 \iint_{R\cap R_M} e^{-(x^2+y^2)} x^{2m-1} y^{2n-1}\, dA,$$

if this limit exists, in which case the limit will be the same whatever the choice of R_M.

Let R_M be the square $0 \leqslant x \leqslant M, 0 \leqslant y \leqslant M$. Then $R \cap R_M = R_M$ and

$$4 \iint_{R\cap R_M} e^{-(x^2+y^2)} x^{2m-1} y^{2n-1}\, dA$$

$$= 4 \int_0^M \int_0^M e^{-x^2} x^{2m-1} e^{-y^2} y^{2n-1}\, dx\, dy$$

$$= \left(2\int_0^M e^{-x^2} x^{2m-1}\, dx\right)\left(2\int_0^M e^{-y^2} y^{2n-1}\, dy\right)$$

$$\to \Gamma(m)\,\Gamma(n),$$

as $M \to \infty$. We have used form (i) for the gamma function. Thus I exists and

$$I = \Gamma(m)\,\Gamma(n).$$

If we take for R_M the region (in polar coordinates) $0 \leqslant \theta \leqslant \pi/2$, $0 \leqslant r \leqslant M$, then as $M \to \infty$ we must obtain the same value for I. But in this case, converting to polar coordinates,

$$4 \iint_{R\cap R_M} e^{-(x^2+y^2)} x^{2m-1} y^{2n-1}\, dA$$

$$= 4 \int_0^{\pi/2} \int_0^M e^{-r^2} (r\cos\theta)^{2m-1} (r\sin\theta)^{2n-1}\, r\, dr\, d\theta$$

$$= 4 \int_0^{\pi/2} \int_0^M e^{-r^2} r^{2(m+n)-1} \cos^{2m-1}\theta \sin^{2n-1}\theta\, dr\, d\theta$$

$$= \left(2\int_0^{\pi/2} \cos^{2m-1}\theta \sin^{2n-1}\theta\, d\theta\right)\left(2\int_0^M e^{-r^2} r^{2(m+n)-1}\, dr\right)$$

$$\to B(m, n)\,\Gamma(m+n),$$

as $M \to \infty$, using form (v) for the beta function.

Hence also

$$I = \mathrm{B}(m, n)\,\Gamma(m+n),$$

and we conclude that

$$\mathrm{B}(m, n) = \frac{\Gamma(m)\,\Gamma(n)}{\Gamma(m+n)}.$$

(This implies another proof that $\Gamma(\frac{1}{2}) = \sqrt{\pi}$: put $m = n = \frac{1}{2}$ in the preceding result: $\mathrm{B}(\frac{1}{2}, \frac{1}{2}) = \pi$, using form (v), and $\Gamma(1) = 1$.)

We use this result to prove the following formula:

$$\Gamma(2p) = \frac{2^{2p-1}}{\sqrt{\pi}}\,\Gamma(p)\,\Gamma(p+\tfrac{1}{2}).$$

Proof:

$$\mathrm{B}(p, p) = 2\int_0^{\pi/2} \sin^{2p-1}\theta\,\cos^{2p-1}\theta\,d\theta$$

$$= 2.2^{1-2p}\int_0^{\pi/2} (2\sin\theta\cos\theta)^{2p-1}\,d\theta$$

$$= 2^{2-2p}\int_0^{\pi/2} \sin^{2p-1} 2\theta\,d\theta$$

$$= 2^{1-2p}\int_0^{\pi} \sin^{2p-1}\phi\,d\phi \qquad [\phi = 2\theta]$$

$$= 2^{1-2p}\,.\,2\int_0^{\pi/2} \sin^{2p-1}\phi\,d\phi$$

$$= 2^{1-2p}\,\mathrm{B}(p, \tfrac{1}{2}).$$

Hence,

$$\frac{\Gamma(p)\,\Gamma(p)}{\Gamma(2p)} = 2^{1-2p}\,\frac{\Gamma(p)\,\Gamma(\frac{1}{2})}{\Gamma(p+\frac{1}{2})}$$

and the required result follows on rearranging and putting $\Gamma(\frac{1}{2}) = \sqrt{\pi}$.

In the particular case when $p = \frac{1}{4}$, we get

$$\Gamma(\tfrac{1}{2}) = \frac{2^{-\frac{1}{2}}}{\sqrt{\pi}}\,\Gamma(\tfrac{1}{4})\,\Gamma(\tfrac{3}{4}),$$

or

$$\Gamma(\tfrac{1}{4})\,\Gamma(\tfrac{3}{4}) = \pi\sqrt{2}.$$

This is a special case of the formula:

$$\Gamma(p)\,\Gamma(1-p) = \frac{\pi}{\sin p\pi}, \qquad 0 < p < 1,$$

whose proof we cannot give in this book.

The important formulae of this section are:

(i) $\Gamma(n) = \int_0^\infty e^{-x} x^{n-1}\,dx = 2\int_0^\infty e^{-x^2} x^{2n-1}\,dx, \quad n > 0;$

(ii) $B(m, n) = \int_0^1 x^{m-1}(1-x)^{n-1}\,dx$

$$= 2\int_0^{\pi/2} \sin^{2m-1}\theta \cos^{2n-1}\theta\,d\theta, \quad m > 0, n > 0;$$

(iii) $\Gamma(1) = 1, \Gamma(\tfrac{1}{2}) = \sqrt{\pi};$

(iv) $\Gamma(n+1) = n\,\Gamma(n);$

(v) $B(m, n) = \dfrac{\Gamma(m)\,\Gamma(n)}{\Gamma(m+n)};$

(vi) $\Gamma(2p) = \dfrac{2^{2p-1}}{\sqrt{\pi}}\,\Gamma(p)\,\Gamma(p+\tfrac{1}{2});$

(vii) $\Gamma(p)\,\Gamma(1-p) = \dfrac{\pi}{\sin p\pi}.$

Expressions involving beta functions may always be given in terms of gamma functions. By virtue of the recurrence formula, we need only know values of $\Gamma(n)$ for n between 0 and 1. Extensive tables have been constructed for $\Gamma(n)$, $0 < n \leqslant 1$.

Example 1. Show that:

$$\int_0^a \int_0^{a-x} \int_0^{a-x-y} x^{l-1} y^{m-1} z^{n-1} f(x+y+z)\,dz\,dy\,dx$$

$$= \frac{\Gamma(l)\,\Gamma(m)\,\Gamma(n)}{\Gamma(l+m+n)} \int_0^a f(u)\,u^{l+m+n-1}\,du,$$

where $l > 0$, $m > 0$, $n > 0$, assuming the function $f(x)$ to be such that the integrals converge.

Solution: The transformation we shall make use of may be written equivalently in either of the following three ways:

$$x = uvw, \qquad x + y = uv, \qquad x + y + z = u;$$

$$x = uvw, \qquad y = uv(1 - w), \qquad z = u(1 - v);$$

$$u = x + y + z, \qquad v = \frac{x + y}{x + y + z}, \qquad w = \frac{x}{x + y}.$$

The Jacobian of the transformation is

$$\frac{\partial(x, y, z)}{\partial(u, v, w)} = \begin{vmatrix} vw & uw & uv \\ v(1 - w) & u(1 - w) & -uv \\ 1 - v & -u & 0 \end{vmatrix} = -u^2v.$$

The region of integration in xyz-space is the tetrahedron bounded by the planes $x = 0$, $y = 0$, $z = 0$, $x + y + z = a$. Under the given transformation, these become respectively the planes $w = 0$, $w = 1$, $v = 1$, and $u = a$ in uvw-space. (This is best seen using the third formulation of the transformation.) These planes do not enclose a finite region in uvw-space; the trouble can be tracked down to the vanishing of the Jacobian. We see that $u^2v = (x + y)(x + y + z) = 0$ whenever x and y are both zero (neither can ever be negative so this is the only way their sum can be zero) so the Jacobian vanishes along the z-axis, part of which forms one edge of the tetrahedron. We therefore exclude this edge by means of the plane $x + y = \epsilon$, where $\epsilon > 0$ (see Fig. 34(a)). This plane is transformed into the hyperbolic cylinder $uv = \epsilon$. Fig. 34(b) now shows the resulting solid in uvw-space. It is

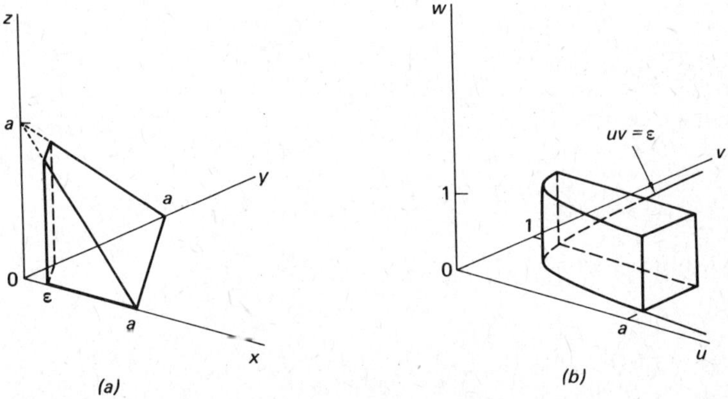

Fig. 34

then easy to see, letting ϵ tend to zero, that the tetrahedron is mapped into the prism defined by $0 \leqslant u \leqslant a, 0 \leqslant v \leqslant 1, 0 \leqslant w \leqslant 1$.

Carrying out the transformation, we thus have

$$\int_0^a \int_0^{a-x} \int_0^{a-x-y} x^{l-1} y^{m-1} z^{n-1} \mathrm{f}(x+y+z)\, \mathrm{d}z\, \mathrm{d}y\, \mathrm{d}x$$

$$= \int_0^a \int_0^1 \int_0^1 \{uvw\}^{l-1} \{uv(1-w)\}^{m-1} \{u(1-v)\}^{n-1} \times$$

$$\mathrm{f}(u)\, u^2 v\, \mathrm{d}w\, \mathrm{d}v\, \mathrm{d}u$$

$$= \int_0^a u^{l+m+n-1} \mathrm{f}(u)\, \mathrm{d}u \,.\, \int_0^1 v^{l+m-1}(1-v)^{n-1}\, \mathrm{d}v \,.\, \int_0^1 w^{l-1} \times$$

$$(1-w)^{m-1}\, \mathrm{d}w$$

$$= \mathrm{B}(l+m, n)\, \mathrm{B}(l, m) \int_0^a u^{l+m+n-1} \mathrm{f}(u)\, \mathrm{d}u$$

$$= \frac{\Gamma(l+m)\,\Gamma(n)}{\Gamma(l+m+n)} \cdot \frac{\Gamma(l)\,\Gamma(m)}{\Gamma(l+m)} \int_0^a u^{l+m+n-1} \mathrm{f}(u)\, \mathrm{d}u,$$

and the required result follows.

The integral that we have discussed here is called Dirichlet's integral.

Example 2. Find the volume in the first octant bounded by the coordinate planes and the surface $x^4 + y^4 + z^4 = a^4$.

Solution: We shall first put $u = x^2$, $v = y^2$, $w = z^2$ and then use spherical coordinates to integrate over that part of the sphere $u^2 + v^2 + w^2 \leqslant a^4$ which lies in the first octant. Since,

$$\frac{\partial(u, v, w)}{\partial(x, y, z)} = 8\,xyz = 8\,\sqrt{(uvw)},$$

the required volume is:

$$\iiint \mathrm{d}x\, \mathrm{d}y\, \mathrm{d}z = \iiint \frac{1}{8\,\sqrt{(uvw)}}\, \mathrm{d}u\, \mathrm{d}v\, \mathrm{d}w$$

$$= \tfrac{1}{8} \int_0^{\pi/2} \int_0^{\pi/2} \int_0^{a^2} \frac{\rho^2 \sin\phi\, \mathrm{d}\rho\, \mathrm{d}\phi\, \mathrm{d}\theta}{\sqrt{(\rho \sin\phi \cos\theta \,.\, \rho \sin\phi \sin\theta \,.\, \rho\cos\phi)}}$$

$$= \tfrac{1}{8}\int_0^{\pi/2}\int_0^{\pi/2}\int_0^{a^2} \rho^{\frac{1}{2}} \cos^{-\frac{1}{2}}\phi \sin^{-\frac{1}{2}}\theta \cos^{-\frac{1}{2}}\theta \, d\rho \, d\phi \, d\theta$$

$$= \frac{1}{32}\cdot\left[\frac{2}{3}\rho^{3/2}\right]_0^{a^2} . 2\int_0^{\pi/2} \cos^{-\frac{1}{2}}\phi \, d\phi \,.\, 2\int_0^{\pi/2} \sin^{-\frac{1}{2}}\theta \cos^{-\frac{1}{2}}\theta \, d\theta$$

$$= \frac{2}{96} a^3 \,\mathrm{B}(\tfrac{1}{4},\tfrac{1}{2})\,\mathrm{B}(\tfrac{1}{4},\tfrac{1}{4})$$

$$= \frac{a^3}{48}\,\frac{\Gamma(\frac{1}{4})\,\Gamma(\frac{1}{2})}{\Gamma(\frac{3}{4})}\,\frac{\Gamma(\frac{1}{4})\,\Gamma(\frac{1}{4})}{\Gamma(\frac{1}{2})}\cdot\frac{\Gamma(\frac{1}{4})}{\Gamma(\frac{1}{4})}$$

$$= \frac{a^3\{\Gamma(\frac{1}{4})\}^4}{48\pi\sqrt{2}},$$

since $\Gamma(\frac{3}{4})\,\Gamma(\frac{1}{4}) = \pi\sqrt{2}$.

EXERCISES 6b

(1) Compute:

$$\Gamma\left(\frac{7}{2}\right), \Gamma\left(\frac{9}{2}\right), \Gamma(5), \mathrm{B}\left(2,\frac{3}{2}\right), \Gamma\left(\frac{1}{6}\right)\Gamma\left(\frac{5}{6}\right),$$

$$\Gamma\left(\frac{1}{3}\right)\Gamma\left(\frac{2}{3}\right), \mathrm{B}\left(\frac{1}{6},\frac{5}{6}\right).$$

(2) Show that the area in the first quadrant bounded by the coordinate axes and the curve $x^4 + y^4 = a^4$ is $a^2\{\Gamma(\frac{1}{4})\}^2/8\sqrt{\pi}$.

(3) Assuming the integrals to converge, show that

$$\int_0^a\int_0^{a-x} x^{m-1}y^{n-1}\,\mathrm{f}(x+y)\,dy\,dx = \mathrm{B}(m,n)\int_0^a \mathrm{f}(u)\,u^{m+n-1}\,du,$$

where $m > 0$ and $n > 0$.

(4) Evaluate:

$$\int_0^1\int_0^{1-x} (xy)^{-3/4}\,dy\,dx,$$

expressing the answer in terms of a single gamma function.

(5) By means of the substitution $u = x^3 + y^3$, $v = y^3$, or otherwise, show that

$$\iint_R x^2 y^3 \sqrt{(1 - x^3 - y^3)}\, \mathrm{d}A = \frac{2}{935\sqrt{\pi}}\, \Gamma(\tfrac{1}{3})\, \Gamma(\tfrac{1}{6}),$$

where the region R is defined by $x \geqslant 0, y \geqslant 0, x^3 + y^3 \leqslant 1$.

(6) Show that:

(i) $$\iiint (x + y + z)^n\, xyz\, \mathrm{d}V = \frac{1}{5!\,(n + 6)}, \qquad n \neq -6,$$

(ii) $$\iiint \frac{x^{l-1} y^{m-1} z^{n-1}}{(x + y + z)^{l+m+n-1}}\, \mathrm{d}V = \frac{\Gamma(l)\,\Gamma(m)\,\Gamma(n)}{\Gamma(l + m + n)}, \qquad l > 0, m > 0, n > 0$$

where the integrations are over the tetrahedron with vertices (0, 0, 0), (0, 0, 1), (0, 1, 0), (1, 0, 0).

(7) Show that, if $l > 0, m > 0, n > 0$,

$$\iiint_S x^{2l-1} y^{2m-1} z^{2n-1}\, \mathrm{d}V = \frac{a^{2l} b^{2m} c^{2n}}{8} \frac{\Gamma(l)\,\Gamma(m)\,\Gamma(n)}{\Gamma(l + m + n + 1)},$$

where S is the interior of that part of the ellipsoid $(x^2/a^2) + (y^2/b^2) + (z^2/c^2) = 1$ lying in the positive octant.

Multiple integrals of higher dimension

Our work so far has solely concerned double and triple integrals as these are the ones occurring most commonly in physical applications. We will now indicate the definition of a multiple integral in n dimensions. Such integrals do have physical applications (for example, various two-body problems in three dimensions require six-fold integrals) but the application we shall give will be one of interest in the advanced theory of statistics. It is the problem of determining the 'volume' of an n-dimensional 'sphere'.

Let $\mathrm{f}(x_1, x_2, \ldots, x_n)$ be a function of n independent variables $x_1, x_2, \ldots, x_n$ defined in some region T of n-dimensional space. As in two and three dimensions, T may presumably be defined by a set of n inequalities. Some concept of the 'volume' (better called content or measure) of T is required. If T is a 'rectangular' region, defined by the inequalities $a_1 \leqslant x_1 \leqslant b_1, a_2 \leqslant x_2 \leqslant b_2, \ldots, a_n \leqslant x_n \leqslant b_n$, then its content is defined as $(b_1 - a_1)(b_2 - a_2) \ldots (b_n - a_n)$.

Restricting ourselves to such regions, the n-fold multiple integral of $f(x_1, x_2, \ldots, x_n)$ over T would be written:

$$\iint \cdots \int_T f(x_1, x_2, \ldots, x_n)\, dx_1\, dx_2 \ldots dx_n,$$

with corresponding repeated integral:

$$\int_{a_n}^{b_n} \cdots \int_{a_2}^{b_2} \int_{a_1}^{b_1} f(x_1, x_2, \ldots, x_n)\, dx_1\, dx_2 \ldots dx_n.$$

The multiple integral is often written more simply as:

$$\int_T f(x_1, x_2, \ldots, x_n)\, d(x_1, x_2, \ldots, x_n).$$

and this suggests the convenient vector notation:

$$\int_T f(\mathbf{x})\, d\mathbf{x},$$

where $\mathbf{x}$ is the n-dimensional vector $(x_1, x_2, \ldots, x_n)$.

The formula for change of variables can also be generalised in straightforward fashion. If the transformation is given by the equations

$$x_1 = x_1(u_1, u_2, \ldots, u_n),$$
$$x_2 = x_2(u_1, u_2, \ldots, u_n),$$
$$\ldots,$$
$$x_n = x_n(u_1, u_2, \ldots, u_n),$$

and the Jacobian $\dfrac{\partial(x_1, x_2, \ldots, x_n)}{\partial(u_1, u_2, \ldots, u_n)}$ is not zero anywhere in T (the Jacobian is an $n \times n$ determinant with a similar construction to the two- and three-dimensional Jacobians) then,

$$\iint \cdots \int_T f(x_1, x_2, \ldots, x_n)\, dx_1\, dx_2 \ldots dx_n$$

$$= \iint \cdots \int_{T'} g(u_1, u_2, \ldots, u_n) \left| \frac{\partial(x_1, x_2, \ldots, x_n)}{\partial(u_1, u_2, \ldots, u_n)} \right| du_1\, du_2 \ldots du_n$$

where T' is the region in $u_1 u_2 \ldots u_n$-space corresponding to the region T

in $x_1x_2 \ldots x_n$-space. The vector notation is particularly convenient:

$$\int_T \mathbf{f(x)}\,d\mathbf{x} = \int_{T'} \mathbf{g(u)} \left|\frac{\partial \mathbf{x}}{\partial \mathbf{u}}\right| d\mathbf{u}.$$

Our application is to find the content of the n-dimensional region T defined by:

$$x_1^2 + x_2^2 + \ldots + x_n^2 \leqslant 1.$$

If $n = 2$, this is a circle of radius 1; if $n = 3$, a sphere of radius 1. T will be referred to as the unit n-ball. We shall give two methods, both of which use a mapping to transform T into a rectangular region.

In the first method, we put

$$\begin{aligned} x_1 &= \cos\theta_1, \\ x_2 &= \sin\theta_1 \cos\theta_2, \\ x_3 &= \sin\theta_1 \sin\theta_2 \cos\theta_3, \\ &\ldots, \\ x_n &= \sin\theta_1 \sin\theta_2 \sin\theta_3 \ldots \sin\theta_{n-1} \cos\theta_n. \end{aligned}$$

Notice that

$$\begin{aligned} x_1^2 &= \cos^2\theta_1 = 1 - \sin^2\theta_1, \\ x_1^2 + x_2^2 &= 1 - \sin^2\theta_1 + \sin^2\theta_1\cos^2\theta_2 = 1 - \sin^2\theta_1 \sin^2\theta_2, \\ x_1^2 + x_2^2 + x_3^2 &= 1 - \sin^2\theta_1\sin^2\theta_2 + \sin^2\theta_1\sin^2\theta_2\cos^2\theta_3 \\ &= 1 - \sin^2\theta_1\sin^2\theta_2\sin^2\theta_3, \end{aligned}$$

and in general,

$$x_1^2 + x_2^2 + \ldots + x_n^2 = 1 - \sin^2\theta_1 \sin^2\theta_2 \ldots \sin^2\theta_n.$$

We can then see that this transformation sets up a one-to-one correspondence between the region T in $x_1x_2 \ldots x_n$-space and the region T' in $\theta_1\theta_2 \ldots \theta_n$-space where T' is the rectangular region $0 \leqslant \theta_i \leqslant \pi$ for $i = 1, 2, \ldots n$. The Jacobian of the transformation is easily determined. It is:

$$\frac{\partial(x_1, x_2, \ldots, x_n)}{\partial(\theta_1, \theta_2, \ldots, \theta_n)} = \begin{vmatrix} \dfrac{\partial x_1}{\partial\theta_1} & \dfrac{\partial x_1}{\partial\theta_2} \cdots & \dfrac{\partial x_1}{\partial\theta_n} \\ \dfrac{\partial x_2}{\partial\theta_1} & \dfrac{\partial x_2}{\partial\theta_2} \cdots & \dfrac{\partial x_2}{\partial\theta_n} \\ \cdots & \cdots & \cdots \\ \dfrac{\partial x_n}{\partial\theta_1} & \dfrac{\partial x_n}{\partial\theta_2} \cdots & \dfrac{\partial x_n}{\partial\theta_n} \end{vmatrix}$$

$$= \begin{vmatrix} -\sin\theta_1 & 0 & \dots \\ \cos\theta_1\cos\theta_2 & -\sin\theta_1\sin\theta_2 & \dots \\ \dots & \dots & \dots \end{vmatrix}$$

and the value of such a determinant is just the product of the diagonal terms, i.e.

$$(-\sin\theta_1)(-\sin\theta_1\sin\theta_2)(-\sin\theta_1\sin\theta_2\sin\theta_3)\dots$$
$$(-\sin\theta_1\sin\theta_2\dots\sin\theta_n)$$
$$=(-1)^n\sin^n\theta_1\sin^{n-1}\theta_2\sin^{n-2}\theta_3\dots\sin^2\theta_{n-1}\sin\theta_n.$$

Hence the required content of the unit n-ball is:

$$\iint\dots\int_T \mathrm{d}x_1\,\mathrm{d}x_2\dots\mathrm{d}x_n$$

$$=\iint\dots\int_{T'}\left|\frac{\partial(x_1,\dots,x_n)}{\partial(\theta_1,\dots,\theta_n)}\right|\mathrm{d}\theta_1\,\mathrm{d}\theta_2\dots\mathrm{d}\theta_n$$

$$=\int_0^\pi\int_0^\pi\dots\int_0^\pi \sin^n\theta_1\sin^{n-1}\theta_2\dots\sin\theta_n\,\mathrm{d}\theta_1\,\mathrm{d}\theta_2\dots\mathrm{d}\theta_n$$

$$=2\int_0^{\pi/2}\sin^n\theta_1\,\mathrm{d}\theta_1\,.\,2\int_0^{\pi/2}\sin^{n-1}\theta_2\,\mathrm{d}\theta_2\dots2\int_0^{\pi/2}\sin\theta_n\,\mathrm{d}\theta_n$$

$$=\mathrm{B}\left(\frac{n+1}{2},\frac{1}{2}\right)\mathrm{B}\left(\frac{n}{2},\frac{1}{2}\right)\mathrm{B}\left(\frac{n-1}{2},\frac{1}{2}\right)\dots\mathrm{B}\left(\frac{3}{2},\frac{1}{2}\right)\mathrm{B}\left(\frac{2}{2},\frac{1}{2}\right)$$

$$=\frac{\Gamma[(n+1)/2]\,\Gamma(\frac{1}{2})}{\Gamma[(n+2)/2]}\cdot\frac{\Gamma(n/2)\,\Gamma(\frac{1}{2})}{\Gamma[(n+1)/2]}\cdot\frac{\Gamma[(n-1)/2]\,\Gamma(\frac{1}{2})}{\Gamma(n/2)}\dots$$
$$\frac{\Gamma(2/2)\,\Gamma(\frac{1}{2})}{\Gamma(3/2)}$$

$$=\frac{[\Gamma(\frac{1}{2})]^n}{\Gamma[(n+2)/2]}$$

$$=\frac{\pi^{n/2}}{\Gamma[(n/2)+1]},$$

since $\Gamma(\frac{1}{2})=\sqrt{\pi}$ and $\Gamma(1)=1$.

The second method uses a natural generalisation of spherical coordinates. We put:

$$x_1 = \rho \cos \phi_1,$$
$$x_2 = \rho \sin \phi_1 \cos \phi_2,$$
$$x_3 = \rho \sin \phi_1 \sin \phi_2 \cos \phi_3,$$
$$\ldots,$$
$$x_{n-1} = \rho \sin \phi_1 \sin \phi_2 \sin \phi_3 \ldots \sin \phi_{n-2} \cos \phi_{n-1},$$
$$x_n = \rho \sin \phi_1 \sin \phi_2 \sin \phi_3 \ldots \sin \phi_{n-2} \sin \phi_{n-1}.$$

The unit n-ball is mapped onto the rectangular region determined by the inequalities

$$0 \leqslant \rho \leqslant 1, \quad 0 \leqslant \phi_i \leqslant \pi \quad \text{for} \quad i = 1, 2, \ldots, n-2,$$

and

$$0 \leqslant \phi_{n-1} \leqslant 2\pi.$$

The Jacobian of this transformation is a little more difficult to compute. It can be shown, using mathematical induction, that:

$$\frac{\partial(x_1, x_2, \ldots, x_n)}{\partial(\rho, \phi_1, \ldots, \phi_{n-1})} = \rho^{n-1} \sin^{n-2} \phi_1 \sin^{n-3} \phi_2 \ldots \sin^2 \phi_{n-3} \sin \phi_{n-2}.$$

The integration then leads, in a similar manner, to the same result, of course.

Since n is always a positive integer, we may write our result as follows, distinguishing even and odd values of n. The content of the unit n-ball is:

$$\frac{\pi^{n/2}}{\Gamma[(n/2)+1]};$$

that is,

$$\pi^k/k!, \qquad \text{when } n = 2k,$$

$$\frac{2^{k+1}\pi^k}{1.3.5.\ldots.(2k+1)}, \qquad \text{when } n = 2k+1.$$

Here we have put $\Gamma(k+1) = k!$ and $\Gamma(k + \frac{1}{2} + 1) = (k + \frac{1}{2})\,\Gamma(k - \frac{1}{2} + 1) = (k + \frac{1}{2})(k - \frac{1}{2})\,\Gamma[k - (3/2) + 1] = \ldots = (k + \frac{1}{2})(k - \frac{1}{2})\,[k - (3/2)] \ldots \frac{1}{2}\,\Gamma(\frac{1}{2})$.

The result should be checked with the well-known results corresponding to $n = 1$, 2 and 3.

It should be clear that for an n-ball of radius a, given by $x_1^2 + x_2^2 + \ldots + x_n^2 \leqslant a^2$, the content is:

$$\frac{\pi^{n/2} a^n}{\Gamma[(n/2) + 1]} .$$

7

Applications of Multiple Integrals

Our approach in previous chapters has stressed heavily the use of double and triple integrals in finding areas and volumes of two- and three-dimensional regions, respectively. These are only two of many applications of multiple integrals. Many of these applications, for example to potential theory, fluid motion and electricity and magnetism, are best handled using the methods of vector analysis. The applications which we shall describe here do not make use of these methods, but some knowledge of physical principles will be called on. Our final application, to multivariate distributions in statistics, does not require any prior knowledge.

Surface area

In the theory of the definite integral, one common application is to find the length of an arc of a curve: if the curve has equation $y = f(x)$, then the length of the arc between $x = a$ and $x = b$ is, under suitable conditions,

$$\int_a^b \sqrt{\{1 + [f'(x)]^2\}}\, dx.$$

This may also be written as:

$$\int_I \sqrt{\{1 + [f'(x)]^2\}}\, dx,$$

where I is the projection of the given arc on the x-axis.

We shall now justify an entirely analogous formula for the area of a curved surface. Suppose the equation of the surface under question is

$z = f(x, y)$ and that it lies over the region R in the xy-plane. Then, under suitable conditions, its surface area is

$$\iint_R \sqrt{\{1 + [f_x(x, y)]^2 + [f_y(x, y)]^2\}}\, dA.$$

Our approach to the problem is somewhat akin to the method used in proving the rule for change of variables in Chapter 5. Choose a rectangle $KLMN$ lying in R with vertices $K(x, y)$, $L(x + \Delta x, y)$, $M(x + \Delta x, y + \Delta y)$, $N(x, y + \Delta y)$. Then $KLMN$ has area $\Delta x \Delta y$ and is one element in a Cartesian partitioning of R (see Fig. 35). On the surface there are the corresponding points $K'x, y, f(x, y)$, $L'(x + \Delta x, y, f(x + \Delta x, y))$, $M'(x + \Delta x, y + \Delta y, f(x + \Delta x, y + \Delta y))$ and $N'(x, y + \Delta y, f(x, y + \Delta y))$.

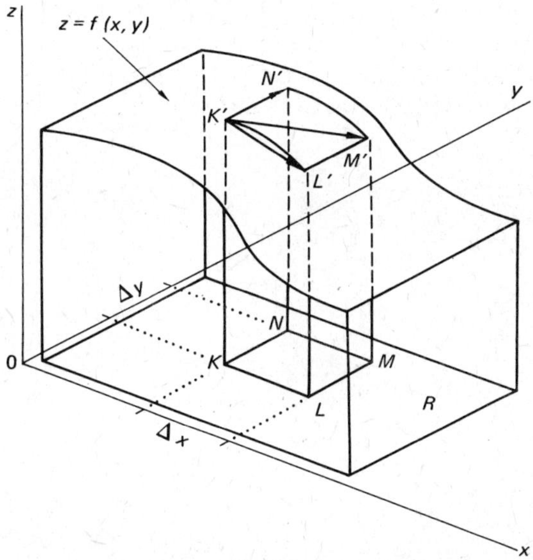

Fig. 35

By the same device used in Chapter 5, we may write these points as

$K'(x, y, f)$,

$L'(x + \Delta x, y, f + f_x \Delta x)$,

$M'(x + \Delta x, y + \Delta y, f + f_x \Delta x + f_y \Delta y)$,

$N'(x, y + \Delta y, f + f_y \Delta y)$,

approximately. But then the vectors $\mathbf{K'L'}$ and $\mathbf{N'M'}$ are equal (both are $\Delta x\, \mathbf{i} + f_x \Delta x\, \mathbf{k}$) and $\mathbf{L'M'}$ and $\mathbf{K'N'}$ are equal (both are $\Delta y\, \mathbf{j} + f_y \Delta y\, \mathbf{k}$), so our approximation amounts to taking $K'L'M'N'$ to be a parallelogram (lying in fact in the tangent plane at K').

For small Δx and Δy the area of this parallelogram is thus approximately the area of that part of the curved surface lying over $KLMN$. Since

$$K'L' \times K'N' = \begin{vmatrix} \mathbf{i} & \mathbf{j} & \mathbf{k} \\ \Delta x & 0 & f_x \Delta x \\ 0 & \Delta y & f_y \Delta y \end{vmatrix}$$

$$= -f_x \Delta x \Delta y\, \mathbf{i} - f_y \Delta x \Delta y\, \mathbf{j} + \Delta x \Delta y\, \mathbf{k},$$

the area of the parallelogram is

$$|K'L' \times K'N'| = \sqrt{\{1 + f_x^2 + f_y^2\}}\, \Delta x \Delta y.$$

Summing this quantity over all rectangles in the partition of R and letting Δx and Δy tend to zero, we obtain the desired expression for the surface area.

If the equation of the surface was given in the form $\phi(x, y, z) = 0$, with z still the dependent variable, we have, differentiating partially with respect to x,

$$\frac{\partial \phi}{\partial x} + \frac{\partial \phi}{\partial z}\frac{\partial z}{\partial x} = 0,$$

so that

$$\frac{\partial z}{\partial x} = -\frac{\phi_x}{\phi_z},$$

and similarly

$$\frac{\partial z}{\partial y} = -\frac{\phi_y}{\phi_z}.$$

Then,

$$\sqrt{\{1 + f_x^2 + f_y^2\}} = \sqrt{\left\{1 + \left(\frac{\partial z}{\partial x}\right)^2 + \left(\frac{\partial z}{\partial y}\right)^2\right\}}$$

$$= \sqrt{\left\{1 + \left(\frac{\phi_x}{\phi_z}\right)^2 + \left(\frac{\phi_y}{\phi_z}\right)^2\right\}}$$

$$= \frac{1}{|\phi_z|} \sqrt{\{\phi_x^2 + \phi_y^2 + \phi_z^2\}}$$

Provided $\phi_z \neq 0$, we can then give another expression for the surface area:

$$\iint_R \frac{1}{|\phi_z|} \sqrt{\{\phi_x^2 + \phi_y^2 + \phi_z^2\}}\, dA,$$

where again R is the projection of the surface on the xy-plane.

There are other corresponding alternative expressions for the surface area, obtained by projecting the surface on either of the other two coordinate planes.

It is often convenient to think of an element of the surface (such as the piece $K'L'M'N'$ in Fig. 35) as having area ΔS, say. Then the total surface area can be written down simply as:

$$\iint_R \mathrm{d}S,$$

where R is the projection of the surface on (for the moment) the xy-plane. This formula is analogous with that given in Chapter 2 for the area of a plane region and is of course a generalisation of it. We have shown above that in computing a surface area we must first put

$$\mathrm{d}S = \sqrt{\left\{1 + \left(\frac{\partial z}{\partial x}\right)^2 + \left(\frac{\partial z}{\partial y}\right)^2\right\}}\ \mathrm{d}A$$

and then integrate over R. If it is necessary to find the area of a surface by projecting it on the yz-plane say, then $\iint_R \mathrm{d}S$ still serves as the required formula, but we must understand that now the equation of the surface needs to be given in the form $x = \mathrm{g}(y, z)$, R is the projection of the surface on this plane, and

$$\mathrm{d}S = \sqrt{\left\{1 + \left(\frac{\partial x}{\partial y}\right)^2 + \left(\frac{\partial x}{\partial z}\right)^2\right\}}\ \mathrm{d}A.$$

Example. Prove that the area of the portion of the sphere $x^2 + y^2 + (z - c)^2 = c^2$ contained within the paraboloid $x^2 + y^2 = 2az$ $(c \geqslant 2a > 0)$ is $4\pi ac$.

Solution: As Fig. 36 shows, we require the surface area of a spherical cap. The projection of this cap on the xy-plane is the circle under the base of the cap. Eliminating z from the two equations,

$$x^2 + y^2 + \left(\frac{x^2 + y^2}{2a} - c\right)^2 = c^2,$$

from which

$$x^2 + y^2 = 4a(c - a),$$

so the circle has radius $2\sqrt{\{a(c - a)\}}$.

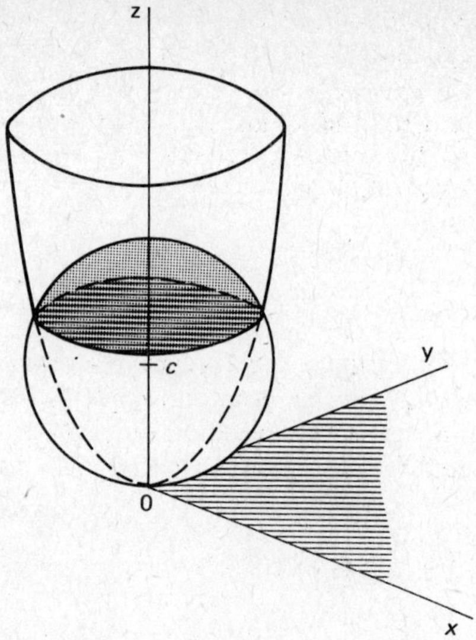

Fig. 36

The equation of the upper half of the sphere is

$$z = c + \sqrt{(c^2 - x^2 - y^2)}.$$

$$\frac{\partial z}{\partial x} = \frac{-x}{\sqrt{(c^2 - x^2 - y^2)}}$$

and

$$\frac{\partial z}{\partial y} = \frac{-y}{\sqrt{(c^2 - x^2 - y^2)}}.$$

The required surface area is thus:

$$\iint dS = \iint \sqrt{\left\{1 + \left(\frac{\partial z}{\partial x}\right)^2 + \left(\frac{\partial z}{\partial y}\right)^2\right\}} \, dA$$

$$= \iint \sqrt{\left\{1 + \frac{x^2}{c^2 - x^2 - y^2} + \frac{y^2}{c^2 - x^2 - y^2}\right\}} \, dA$$

$$= \iint \frac{c}{\sqrt{(c^2 - x^2 - y^2)}} \, dA,$$

the double integrals being over the circle obtained above. Changing to polar coordinates, the latter integral is

$$c\int_0^{2\pi}\int_0^{2\sqrt{\{a(c-a)\}}}\frac{r\,dr\,d\theta}{\sqrt{(c^2-r^2)}}=2\pi c\Big[-\sqrt{(c^2-r^2)}\Big]_0^{2\sqrt{\{a(c-a)\}}}$$

$$= 2\pi c(c-\sqrt{(c^2-4ac+4a^2)})$$

$$= 2\pi c(c-[c-2a])$$

$$= 4\pi ac.$$

Notice that when $a = \frac{1}{2}c$, the cap becomes the full upper hemisphere, with surface area $4\pi \,.\, \frac{1}{2} \,.\, c \,.\, c = 2\pi c^2$ and so we obtain the well-known $4\pi c^2$ for the surface area of a sphere with radius c. (This result is easily obtained in a more direct fashion.)

EXERCISES 7a

(1) Prove that the area of the surface $z^2 = 2xy$ included between the planes $x = 0, x = a, y = 0, y = b$ $(a > 0, b > 0)$ is $\frac{2}{3}\sqrt{(2ab)}(a + b)$.

(2) Find the area of that part of the cone $z^2 = x^2 + y^2$ which lies inside the cylinder $x^2 + y^2 = 2x$.

(3) Find the area of the cylinder $x^2 + y^2 = a^2$ in the first octant included between the planes $z = 0$ and $z = mx$.

(4) Show that the area of the surface of the sphere $x^2 + y^2 + z^2 = a^2$ included within the cylinder $x^2 + y^2 = ax$ is $(2\pi - 4)a^2$.

(5) Find the area of the part of the paraboloid $x^2 + y^2 = 2z$ which lies inside the cylinder $(x^2 + y^2)^2 = x^2 - y^2$.

(6) Show that the surface $z^2 = 2xy$ intersects the sphere $x^2 + y^2 + z^2 = a^2$ in the planes $x + y + a = 0$ and $x + y - a = 0$ and hence find the area of that part of this surface lying inside the sphere.

(7) Prove that the area of the portion of the sphere $x^2 + y^2 + (z - c)^2 = c^2$ contained within the paraboloid $(x^2/a) + (y^2/b) = 2z$ is $4\pi c\sqrt{(ab)}$.

(8) Show that the area of the part of the cone $z^2 = 2xy$ which lies within the cylinder $x^2 + y^2 = 2ax$ is $2\pi a^2$ [Note: gamma functions will be required.]

Centre of mass and moments of inertia

The moment (or first moment) of a particle of mass m with respect to a point distant x from it is the product mx. For a system of n particles, of masses $m_1, m_2, \ldots, m_n$, strung along the x-axis at distances $x_1, x_2, \ldots, x_n$ from the origin, their combined moment about the origin is just the sum of their individual moments; that is,

$$\sum_{i=1}^{n} m_i x_i.$$

If the particles are situated in a plane, at the points (x_1, y_1), (x_2, y_2), $\ldots$, (x_n, y_n), we define their moments with respect to the x- and y-axes as

$$\sum_{i=1}^{n} m_i y_i$$

and

$$\sum_{i=1}^{n} m_i x_i,$$

respectively. (Notice that the ith particle is distant y_i from the x-axis and x_i from the y-axis.)

Finally, if the particles are situated in space, at the points (x_1, y_1, z_1), (x_2, y_2, z_2), $\ldots$, (x_n, y_n, z_n), their moments with respect to the xy-plane, the zx-plane and the yz-plane are

$$\sum_{i=1}^{n} m_i z_i, \quad \sum_{i=1}^{n} m_i y_i$$

and

$$\sum_{i=1}^{n} m_i x_i,$$

respectively. (Notice, for example, that z_i is the distance of the ith particle from the xy-plane.)

The centre of mass (which term we use interchangeably here with centre of gravity) of a system of particles is the point at which a particle whose mass is the total mass of the system has the same moments as the total moments of the system. Let M be the total mass of the n particles discussed above; that is,

$$M = \sum_{i=1}^{n} m_i.$$

In the one-dimensional case, a particle of this mass at a distance $\bar{x}$ from the origin will have moment $M\bar{x}$, and if this is to equal the total moment of the system we require:

$$M\bar{x} = \sum_{i=1}^{n} m_i x_i.$$

Thus,

$$\bar{x} = \frac{\Sigma m_i x_i}{M}$$

is the centre of mass in this case. Similarly in two dimensions the centre of mass will be the point $(\bar{x}, \bar{y})$ at which $Mx = \Sigma m_i x_i$ and $My = \Sigma m_i y_i$, so that

$$\bar{x} = \frac{\Sigma m_i x_i}{M}, \quad \bar{y} = \frac{\Sigma m_i y_i}{M}.$$

Finally, in the case of the n particles in space, their centre of mass is the point $(\bar{x}, \bar{y}, \bar{z})$ where

$$\bar{x} = \frac{\Sigma m_i x_i}{M}, \quad \bar{y} = \frac{\Sigma m_i y_i}{M}, \quad \bar{z} = \frac{\Sigma m_i z_i}{M},$$

with i ranging from 1 to n as usual in the summations.

We now move on to consider a continuous distribution of mass. This may take the form of a (one-dimensional) wire, a (two-dimensional) lamina or a (three-dimensional) solid or of a (three-dimensional) twisted wire or surface.

Suppose we have a solid region S of volume V and mass M. Volume and mass are related by density which we shall assume is not necessarily constant throughout the region. The density may therefore be different at different points of S: let $\mu(x, y, z)$ be the density function, giving the density at the point (x, y, z).

In the usual way, partition S in some way, so that we may speak of a typical element of volume ΔV; let (x', y', z') be some point in this element.

If the partition is fine enough, the density of this element is approximately $\mu(x', y', z')$ throughout it and the mass ΔM of the element is given by:

$$\Delta M = \mu(x', y', z')\,\Delta V.$$

$\sum_{\Delta V} \Delta M$ is clearly an approximation to the total mass of S and the approximation is improved as $(\Delta V)_{max}$ is decreased. Thus,

$$M = \lim \sum_{\Delta V} \mu(x', y', z')\,\Delta V.$$

By definition (see Chapter 3), this limit is a triple integral:

$$M = \iiint_S \mu(x, y, z)\,\mathrm{d}V$$

gives the mass of S as the triple integral over S of the density function. It is often convenient to write simply

$$M = \iiint_S \mathrm{d}M.$$

Likewise, we may say that the moment of the element of mass ΔM with respect to the yz-plane is approximately $x'\Delta M$ or $x'\mu(x', y', z')\,\Delta V$. Summing these and taking the limit as $(\Delta V)_{max} \to 0$, we obtain as the first moment of S with respect to the yz-plane

$$\iiint_S x\,\mu(x, y, z)\,\mathrm{d}V$$

or simply

$$\iiint_S x\,\mathrm{d}M.$$

The x-coordinate of the centre of mass of S is then the value $\bar{x}$ for which $M\bar{x} = \iiint_S x\,\mu(x, y, z)\,\mathrm{d}V$. In this way, we determine that the centre of mass of S is the point $(\bar{x}, \bar{y}, \bar{z})$ where

$$\bar{x} = \frac{\iiint_S x\,\mu\,\mathrm{d}V}{M}, \quad \bar{y} = \frac{\iiint_S y\,\mu\,\mathrm{d}V}{M}, \quad \bar{z} = \frac{\iiint_S z\,\mu\,\mathrm{d}V}{M},$$

with $M = \iiint_S \mu\,\mathrm{d}V.$

It is perhaps more convenient to remember this in the form

$$\bar{x} = \frac{\iiint_S x \, dM}{M}, \quad \bar{y} = \frac{\iiint_S y \, dM}{M}, \quad \bar{z} = \frac{\iiint_S z \, dM}{M},$$

for now the analogy with the discrete case of the n particles is more striking and furthermore this analogy extends to any other body for which the centre of mass is required. Thus the centre of mass of a plane region R may be written down immediately as $(\bar{x}, \bar{y})$ where

$$\bar{x} = \frac{\iint_R x \, dM}{M}, \quad \bar{y} = \frac{\iint_R y \, dM}{M},$$

with $M = \iint_R dM$. In calculations, we replace dM by μdA where now $\mu = \mu(x, y)$ is an areal density function.

In the case of a surface $z = f(x, y)$ (which is commonly used to approximate a thin shell) the arguments above and in the preceding section imply that the centre of mass is given by

$$\bar{x} = \frac{\iint_R x \, dM}{M}, \quad \bar{y} = \frac{\iint_R y \, dM}{M}, \quad \bar{z} = \frac{\iint_R z \, dM}{M},$$

with $M = \iint_R dM$, where R is the projection of the surface on the xy-plane. In calculations we would here replace dM by $\mu \, dS$, where $\mu = \mu(x, y, z) = \mu[x, y, f(x, y)]$ is the surface density function. (Alternatively, the surface may be projected on either of the other coordinate planes. There will be corresponding changes in the above formulae.)

When a region has uniform density, its centre of mass is also known as its centroid.

Returning to the n particles in space, their moment of inertia (or second moment) with respect to some line is $\sum_{i=1}^{n} m_i r_i^2$ where r_i is the distance from the ith particle to the line.

This quantity is met in problems in mechanics involving rotational motion. For example, if all of the n particles are rotating about the line with constant angular speed ω, then the kinetic energy of the system is

$$\tfrac{1}{2}\left(\sum_{i=1}^{n} m_i r_i^2\right)\omega^2.$$

The distance from the point (x_i, y_i, z_i) to the x-axis is $\sqrt{(y_i^2 + z_i^2)}$ so the moment of inertia of the particles with respect to the x-axis, denoted by I_X, is given by:

$$I_X = \sum_{i=1}^{n} m_i (y_i^2 + z_i^2).$$

Similarly the moments of inertia with respect to the y- and z-axes are:

$$I_Y = \sum_{i=1}^{n} m_i (z_i^2 + x_i^2), \; I_Z = \sum_{i=1}^{n} m_i (x_i^2 + y_i^2),$$

respectively.

For a solid region S, these equations transform in the same manner as those for centre of mass. We obtain:

$$I_X = \iiint_S (y^2 + z^2)\, dM,$$

$$I_Y = \iiint_S (z^2 + x^2)\, dM,$$

$$I_Z = \iiint_S (x^2 + y^2)\, dM,$$

where here dM may be replaced by $\mu(x, y, z)\, dV$. As before, the equations in this form may be readily altered for other continuous distributions of mass. For a lamina R in the xy-plane, we have

$$I_X = \iint_R y^2\, dM \quad \text{and} \quad I_Y = \iint_R x^2\, dM,$$

with $dM = \mu(x, y)\, dA$, and for a surface $z = f(x, y)$

$$I_X = \iint_R (y^2 + z^2)\, dM, \; I_Y = \iint_R (z^2 + x^2)\, dM,$$

$$I_Z = \iint_R (x^2 + y^2)\, dM,$$

with $dM = \mu(x, y, z)\, dS$, R being the projection of the surface on the xy-plane.

Example 1. Find the centre of mass of:

(i) the hemispherical solid $x^2 + y^2 + z^2 \leqslant 1,\ z \geqslant 0$, and

(ii) the hemispherical surface $x^2 + y^2 + z^2 = 1,\ z \geqslant 0$, where the volume and surface densities, respectively, are taken to equal 1.

Solution: In both (i) and (ii), the symmetry of the regions implies that $\bar{x} = \bar{y} = 0$. This may be verified by computation.

(i) The mass M of the solid is just its volume $\frac{2}{3}\pi$, since its density is 1. Hence, introducing spherical coordinates,

$$M\bar{z} = \iiint z \, \mathrm{d}M$$

$$= \iiint z \, \mathrm{d}V$$

$$= \int_0^{2\pi} \int_0^{\pi/2} \int_0^1 \rho \cos\phi \,.\, \rho^2 \sin\phi \, \mathrm{d}\rho \, \mathrm{d}\phi \, \mathrm{d}\theta$$

$$= \int_0^{2\pi} \mathrm{d}\theta \,.\, \int_0^{\pi/2} \cos\phi \sin\phi \, \mathrm{d}\phi \,.\, \int_0^1 \rho^3 \, \mathrm{d}\rho$$

$$= 2\pi \,.\, \tfrac{1}{2} \,.\, \tfrac{1}{4}$$

$$= \tfrac{1}{4}\pi.$$

Thus,

$$\bar{z} = \frac{\pi/4}{2\pi/3} = \tfrac{3}{8}$$

so the centre of mass of the hemispherical solid is $(0, 0, \frac{3}{8})$.

(ii) The mass M of the surface is just its area 2π, since its density is 1. Writing $z = \sqrt{(1 - x^2 - y^2)}$, we see that

$$\frac{\partial z}{\partial x} = \frac{-x}{\sqrt{(1 - x^2 - y^2)}}, \quad \frac{\partial z}{\partial y} = \frac{-y}{\sqrt{(1 - x^2 - y^2)}},$$

so

$$M\bar{z} = \iint z \, \mathrm{d}M$$

$$= \iint z \, \mathrm{d}S$$

$$= \iint z \sqrt{\left\{1 + \left(\frac{\partial z}{\partial x}\right)^2 + \left(\frac{\partial z}{\partial y}\right)^2\right\}} \mathrm{d}A$$

$$= \iint z \sqrt{\left(1 + \frac{x^2}{1 - x^2 - y^2} + \frac{y^2}{1 - x^2 - y^2}\right)} \, \mathrm{d}A$$

$$= \iint \sqrt{(1 - x^2 - y^2)} \sqrt{\left(\frac{1}{1 - x^2 - y^2}\right)} \, \mathrm{d}A$$

$$= \iint \mathrm{d}A,$$

$$= \pi,$$

since the double integration is over the circle $x^2 + y^2 \leqslant 1$, which has area π. Thus $\bar{z} = \frac{1}{2}$ so the centre of mass of the hemispherical surface is $(0, 0, \frac{1}{2})$.

It is interesting to consider the annular region $(1 - \epsilon)^2 \leqslant x^2 + y^2 + z^2 \leqslant 1$, $z \geqslant 0$, of unit density. This has $\bar{x} = \bar{y} = 0$ and mass $M = \frac{2}{3}\pi - \frac{2}{3}\pi(1 - \epsilon)^3 = \frac{2}{3}\pi(3\epsilon - 3\epsilon^2 + \epsilon^3)$.

$$M\bar{z} = \iiint z \, \mathrm{dV}$$

$$= \int_0^{2\pi} \int_0^{\pi/2} \int_{1-\epsilon}^{1} \rho^3 \cos\phi \sin\phi \, \mathrm{d}\rho \, \mathrm{d}\phi \, \mathrm{d}\theta$$

$$= 2\pi \cdot \tfrac{1}{2} \cdot \tfrac{1}{4} \left[\rho^4\right]_{1-\epsilon}^{1}$$

$$= \tfrac{1}{4}\pi(4\epsilon - 6\epsilon^2 + 4\epsilon^3 - \epsilon^4).$$

Hence,

$$\bar{z} = \frac{\pi\epsilon(4 - 6\epsilon + 4\epsilon^2 - \epsilon^3)/4}{2\pi\epsilon(3 - 3\epsilon + \epsilon^2)/3}$$

$$= \frac{3}{8} \frac{4 - 6\epsilon + 4\epsilon^2 - \epsilon^3}{3 - 3\epsilon + \epsilon^2}.$$

When $\epsilon = 1$, this gives $\bar{z} = \frac{3}{8}$, coinciding with the result in (i), and when $\epsilon \to 0$, we get $\lim \bar{z} = \frac{1}{2}$, coinciding with the result in (ii).

Example 2. For the uniform solid bounded by the cylinder $x^2 + y^2 = a^2$, the paraboloid $z = b(x^2 + y^2)$, $b > 0$, and the plane $z = 0$, find:

(i) the centre of mass, and

(ii) the moment of inertia with respect to the z-axis.

Solution: The solid is shown in Fig. 37. In cylindrical coordinates, the cylinder has equation $r = a$, and the paraboloid has equation $z = br^2$. The solid may be defined by the inequalities $0 \leqslant \theta \leqslant 2\pi$, $0 \leqslant r \leqslant a$, $0 \leqslant z \leqslant br^2$. Let the density be μ, constant.

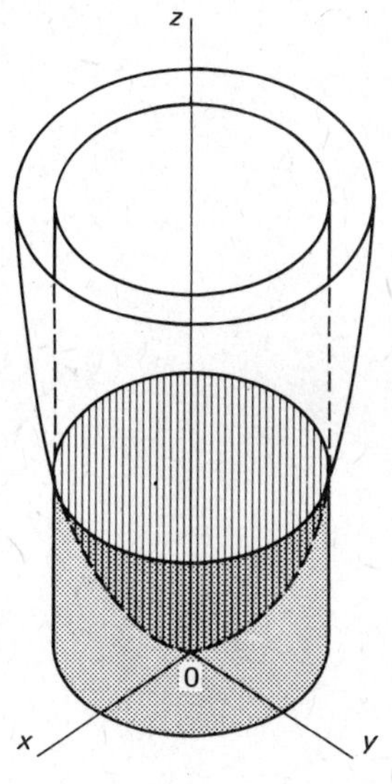

Fig. 37

(i) By symmetry, $\bar{x} = \bar{y} = 0$:

$$z = \frac{\iiint z \,\mathrm{d}M}{\iiint \mathrm{d}M}$$

$$= \frac{\iiint z\,\mu \,\mathrm{d}V}{\iiint \mu \,\mathrm{d}V}$$

$$= \frac{\mu \displaystyle\int_0^{2\pi}\int_0^a\int_0^{br^2} z\, r \,\mathrm{d}z \,\mathrm{d}r \,\mathrm{d}\theta}{\mu \displaystyle\int_0^{2\pi}\int_0^a\int_0^{br^2} r \,\mathrm{d}z \,\mathrm{d}r \,\mathrm{d}\theta}$$

$$= \frac{2\pi \int_0^a r\left[\frac{1}{2} z^2\right]_0^{br^2} dr}{2\pi \int_0^a r\left[z\right]_0^{br^2} dr}$$

$$= \frac{\frac{1}{2} b^2 \int_0^a r^5 \, dr}{b \int_0^a r^3 \, dr}$$

$$= \tfrac{1}{3} a^2 b.$$

The centre of mass is the point $(0, 0, \frac{1}{3} a^2 b)$.

(ii) $I_Z = \iiint (x^2 + y^2) \, dM$

$$= \mu \iiint (x^2 + y^2) \, dV$$

$$= \mu \int_0^{2\pi} \int_0^a \int_0^{br^2} r^3 \, dz \, dr \, d\theta$$

$$= 2\pi\mu b \int_0^a r^5 \, dr$$

$$= \tfrac{1}{3} \pi\mu a^6 b.$$

In (i) we found in passing that the mass of the solid is $M = \frac{1}{2} \pi\mu a^4 b$, so we may write

$$I_Z = \tfrac{2}{3} Ma^2,$$

an expression independent of the density.

Example 3. Find the mass M of a sphere whose density at any point is inversely proportional to the distance of the point from the centre of the sphere and show that its moment of inertia about any diameter is $\frac{1}{3} Ma^2$, where a is the radius of the sphere.

Solution: Let the sphere have its centre at the origin of a spherical coordinate system. Its density at the point (ρ, ϕ, θ) is then k/ρ, where k is the proportionality constant. Now,

$$M = \iiint \mathrm{d}M$$

$$= \iiint (k/\rho)\, \mathrm{d}V$$

$$= k \int_0^{2\pi} \int_0^{\pi} \int_0^{a} \rho \sin\phi \, \mathrm{d}\rho \, \mathrm{d}\phi \, \mathrm{d}\theta$$

$$= k \,.\, 2\pi \,.\, 2 \,.\, \tfrac{1}{2}a^2$$

$$= 2\pi k a^2 .$$

The moment of inertia I will be the same about any diameter. In particular,

$$I_X = \iiint (y^2 + z^2)\, \mathrm{d}M,$$

$$I_Y = \iiint (z^2 + x^2)\, \mathrm{d}M,$$

$$I_Z = \iiint (x^2 + y^2)\, \mathrm{d}M,$$

are all equal, so, adding these three equations,

$$3I = 2\iiint (x^2 + y^2 + z^2)\, \mathrm{d}M$$

$$= 2k \int_0^{2\pi} \int_0^{\pi} \int_0^{a} \rho^2 \,.\, \frac{1}{\rho} \,.\, \rho^2 \sin\phi \, \mathrm{d}\rho \, \mathrm{d}\phi \, \mathrm{d}\theta$$

$$= 2k \,.\, 2\pi \,.\, 2 \,.\, \tfrac{1}{4}a^4$$

$$= 2\pi k a^4 .$$

Hence

$$I = \tfrac{2}{3}\pi k a^4 = \tfrac{1}{3} M a^2 ,$$

on putting $k = M/2\pi a^2$.

EXERCISES 7b

(1) Find the centre of mass of a hemispherical shell having outer radius a and inner radius b if the density (i) is constant, (ii) varies as the square of the distance from the base.

(2) A lamina bounded by the parabolas $y^2 = 4x$, $2x^2 = y$ has a variable density μ given by $\mu(x, y) = 1 + xy$. Show that its mass is unity and find its centre of mass.

(3) Find the centre of mass of the curved surface of the cone $z^2 = x^2 + y^2$, $0 \leqslant z \leqslant h$. Assume constant density.

(4) Determine the moment of inertia of the solid enclosed between the two cylinders $x^2 + y^2 = a^2$ and $x^2 + y^2 = b^2 (a > b)$ and the two planes $z = h$ and $z = -h$, with respect to (i) the z-axis, (ii) the x-axis. Assume unit density.

(5) Find the volume and centre of mass of the uniform region bounded by the parabolic cylinder $z = 4 - x^2$ and the planes $x = 0, y = 0, z = 0, y = 6$.

(6) Find the mass M of the solid, of constant density, bounded by the ellipsoid $(x^2/a^2) + (y^2/b^2) + (z^2/c^2) = 1$, and show that its moment of inertia about the z-axis is $\frac{1}{5} M(a^2 + b^2)$.

(7) A spherical sector may be formed by rotating a sector of a circle about its axis of symmetry. Suppose the circular sector has radius a and angle 2α and that it is symmetrical with respect to the z-axis. Show that the centre of mass of the solid so formed is the point $(0, 0, \frac{3}{8} a (1 + \cos \alpha))$, assuming it to have constant density. What is the centre of mass of the spherical sector if its density at any point is proportional to its distance from the centre?

Find also the centre of mass of the surface of the cap of the spherical sector, assuming constant surface density.

(8) Find the centre of mass of the solid bounded by the paraboloid $z^2 + y^2 = px$ and the plane $x = c$, $p > 0, c > 0$. Assume constant density.

(9) A hemispherical solid is bounded above by the sphere $\rho = R$ and below by the xy-plane, and has density proportional to ρ^{-n}.

(i) For what values of n is the mass finite? Find the mass when it is finite.

(ii) Find the moment of inertia of the solid about the z-axis, stating when it is finite.

(10) A solid region is bounded by the cylinder $x^2 + y^2 = 2ax$ and the planes $z = 0$ and $z = x$. The density at any point of it is proportional to the distance of that point from the xy-plane.

Find an expression for the mass M of the solid and show that its moment of inertia with respect to the z-axis, I_Z, is given by:

$$I_Z = \frac{7}{3} Ma^2.$$

(11) Show that the area of the surface of the paraboloid $x^2 + y^2 = 4z$, below the plane $z = 1$, is $8\pi(2\sqrt{2}-1)/3$. Find its centre of mass, assuming unit surface density, and, in the usual notation, show that $I_Z = 4M\bar{z}$.

Potential of attracting masses

According to Newton's Law of Gravitation, a fixed particle Q of mass m attracts a second particle P of unit mass with a force given by m/D^2, disregarding the gravitational constant, where D is the distance between P and Q.

Placing Q at the origin, for the moment, it is an elementary application of definite integrals to show that the work done by this force in bringing a particle of unit mass from infinity to a point P at a positive distance x from Q along the x-axis is:

$$-\int_{\infty}^{x} \frac{m\,\mathrm{d}D}{D^2} = \frac{m}{x}.$$

This quantity is called the gravitational potential at P due to the particle at Q. In the same way, if Q was now at a fixed point (a, b, c) in space and P at some point (x, y, z), then the potential at P due to Q would be m/D, where $D = \sqrt{\{(x-a)^2 + (y-b)^2 + (z-c)^2\}}$, the distance between P and Q.

For a number of particles $Q_1, Q_2, \ldots, Q_n$ of masses $m_1, m_2, \ldots, m_n$ at the points $(a_1, b_1, c_1), (a_2, b_2, c_2), \ldots, (a_n, b_n, c_n)$, the total potential at the point P is just the sum of the individual potentials:

$$\sum_{i=1}^{n} \frac{m_i}{D_i}, \quad D_i = \sqrt{\{(x-a_i)^2 + (y-b_i)^2 + (z-c_i)^2\}}.$$

We now suppose that instead of the discrete system $Q_1, \ldots, Q_n$, we have a continuous distribution of mass, as in the previous section. The transfer is immediate. For example, the potential at the point $P(x, y, z)$ due to a solid region S is

$$\iiint_S \frac{\mathrm{d}M}{D}, \quad D = \sqrt{\{(x-a)^2 + (y-b)^2 + (z-c)^2\}},$$

where we may replace dM by $\mu(a, b, c)$ da db dc, μ being the density function. Note here, and similarly below, that (a, b, c) is our general point in S so that the integrations are with respect to a, b and c (if Cartesian coordinates are being employed). The potential we calculate by this formula is a function of x, y and z.

The potential at $P(x, y, z)$ due to a surface $z = \mathrm{f}(x, y)$ is

$$\iint_R \frac{\mathrm{d}M}{D}, \quad D = \sqrt{\{(x-a)^2 + (y-b)^2 + (z-c)^2\}},$$

where R is the projection of the surface on the xy-plane and where we may replace dM by $\mu(a, b, c)$ dS.

If the point P lies inside the region of integration, the integral giving the potential at P will be improper (see Chapter 6), but will converge for bounded density functions. This follows from the test implied by Question (4) of Exercises 6a (p. 95).

Example. Find the potential due to a solid sphere at points both outside and inside the sphere. Assume that the sphere has constant density.

Solution: The coordinate system may be chosen so that the sphere has equation $x^2 + y^2 + z^2 = R^2$ and the point has coordinates $(0, 0, z)$, where we may assume $z \geqslant 0$. We are concerned with the two cases:

(i) $z > R$, and

(ii) $z < R$.

Let the density of the solid be μ, constant.

The potential at $(0, 0, z)$ due to the solid sphere is V where

$$V = \iiint \frac{\mathrm{d}M}{D}$$

$$= \iiint \frac{\mu \, \mathrm{d}a \, \mathrm{d}b \, \mathrm{d}c}{\sqrt{\{a^2 + b^2 + (c-z)^2\}}}$$

Change to spherical coordinates: $a = \rho \sin\phi \cos\theta$, $b = \rho \sin\phi \sin\theta$, $c = \rho\cos\phi$, so that

$$V = \mu \int_0^{2\pi} \int_0^{\pi} \int_0^{R} \frac{\rho^2 \sin\phi \, \mathrm{d}\rho \, \mathrm{d}\phi \, \mathrm{d}\theta}{\sqrt{\{\rho^2 \sin^2\phi + (\rho\cos\phi - z)^2\}}}$$

$$= \mu \, . \, 2\pi \int_0^{R} \int_0^{\pi} \frac{\rho^2 \sin\phi \, \mathrm{d}\phi \, \mathrm{d}\rho}{\sqrt{\{\rho^2 - 2\rho z\cos\phi + z^2\}}}$$

In the inner integral, put $t^2 = \rho^2 - 2\rho z \cos\phi + z^2$. Then $t\,dt = \rho z \sin\phi\, d\phi$ and this integral becomes:

$$\frac{\rho}{z}\int_{|\rho - z|}^{\rho + z} \frac{t\,dt}{t} = \frac{\rho}{z}(\rho + z - |\rho - z|), \qquad z > 0.$$

We must now distinguish the two cases.

(i) If $z > R$, then $|\rho - z| = -(\rho - z)$ and

$$V = 2\pi\mu \int_0^R \frac{\rho}{z}.\,2\rho\,d\rho$$

$$= \frac{4\pi R^3}{3z}\mu.$$

(ii) If $0 < z < R$,

$$V = 2\pi\mu\left\{\int_0^z \frac{\rho}{z}.\,2\rho\,d\rho + \int_z^R \frac{\rho}{z}.\,2z\,d\rho\right\}$$

$$= 2\pi\mu\left(\tfrac{2}{3}z^2 + R^2 - z^2\right)$$

$$= 2\pi\mu\left(R^2 - \tfrac{1}{3}z^2\right),$$

while if $z = 0$,

$$V = 2\pi\mu\int_0^R\int_0^\pi \rho\sin\phi\,d\phi\,d\rho$$

$$= 2\pi\mu R^2.$$

Notice that the result in (i) shows that the potential at an external point is the same as if the total mass of the sphere $(4/3)\,\pi R^3\mu$ were concentrated at its centre. Also we see that the potential is continuous at the surface of the sphere: letting z tend to R in both (i) and (ii) gives $V = (4/3)\,\pi R^2\mu$ on the surface.

EXERCISES 7c

(1) A uniform solid spherical shell has radii R' and R ($R' < R$). Find the potential due to it at points outside the outer sphere and inside the inner sphere.

(2) Find the potential due to a uniform spherical surface at points both outside and inside the sphere.

(3) Show that the potential of the solid ellipsoid of revolution

$$\frac{x^2+y^2}{a^2}+\frac{z^2}{b^2}\leqslant 1 \; (b>a)$$

at its centre is

$$\frac{2\pi a^2 b}{\sqrt{(b^2-a^2)}}\log\frac{b+\sqrt{(b^2-a^2)}}{a},$$

when the solid has unit density.

(4) A solid sphere is such that its density at any point is inversely proportional to the distance of that point from the centre. Find the potential due to it at points both inside and outside the sphere.

Multivariate distributions in statistics

First courses in mathematical statistics generally deal in any detail only with distributions of a single random variable. Readers who have done such a course will recognise that the following definitions are quite natural extensions of those they are familiar with.

Suppose a random experiment which is to be repeated many times is such that its outcome at any trial requires more than one number to be given to describe it accurately. For example, if a single die is thrown, the number appearing on the uppermost face is sufficient to describe the result adequately, but if two dice are thrown both numbers may be required. Let $X, Y, \ldots$ be the numbers which serve to describe the outcome of the experiment. These may vary from trial to trial and so are called random variables (or variates). The set of values that $X, Y, \ldots$ can take is called the sample space of the experiment.

For convenience, we will restrict our attention to experiments whose outcomes are describable by three random variables X, Y, Z. The sample space may be some discrete set of points in space or, what is of more interest to us, some solid region. For example, the 'vital statistics' of a group of women would constitute three random variables, which (for argument's sake!) may each range from 10 cm to 500 cm; the sample space would be the cube $10 \leqslant x \leqslant 500$, $10 \leqslant y \leqslant 500$, $10 \leqslant z \leqslant 500$.

We assume here an understanding of the notion of probability. By a distribution of one or more random variables is generally meant a probability distribution function which associates with any subset of the sample space the probability that the outcome of the experiment lies in that subset. Let the sample space of our experiment be some solid region

S. The probability distribution function of this experiment is completely determined by an associated function $f(x, y, z)$, called the frequency function (or probability density function) which must be such that

(i) $f(x, y, z) \geqslant 0$ for all points (x, y, z) in S;

(ii) $\iiint_S f(x, y, z)\, dV = 1.$

Then the probability that our random variables X, Y, Z take values lying in a subset S_1 of S is denoted by $\Pr\{(X, Y, Z) \in S_1\}$ and given by

$$\Pr\{(X, Y, Z) \in S_1\} = \iiint_{S_1} f(x, y, z)\, dV.$$

(The symbol $\in$ is used here as it is in set theory. It may be read as 'is a member of' or 'belongs to'.)

Example. Let:

$$f(x, y, z) = z + 2xy.$$

(i) Show that $f(x, y, z)$ is a frequency function on the sample space $S: 0 \leqslant x \leqslant 1,\ 0 \leqslant y \leqslant 1,\ 0 \leqslant z \leqslant 1$.

(ii) Find $\Pr\{(X, Y, Z) \in S_1\}$ where S_1 is the subset of the cube bounded by the planes $x = 0, y = 0, z = 0, y + z = 1$.

(iii) Find $\Pr\{0 \leqslant X \leqslant 1,\ 0 \leqslant Y \leqslant \frac{1}{2},\ 0 \leqslant Z \leqslant 1\}$.

(iv) Find $\Pr\{X + Y + Z \leqslant 1\}$.

Solution: (Note the alternative methods, in (iii) and (iv), of specifying a subset of the sample space.)

(i) We must verify that:

$$f(x, y, z) \geqslant 0 \text{ in } S,$$

which is clear, and that

$$\iiint_S f(x, y, z)\, dV = 1.$$

Here,

$$\iiint_S f(x, y, z)\, dV = \int_0^1 \int_0^1 \int_0^1 (z + 2xy)\, dz\, dy\, dx$$

$$= \tfrac{1}{2} + 2 \cdot \tfrac{1}{2} \cdot \tfrac{1}{2}$$

$$= 1,$$

as required.

(ii) S_1 may be given by the inequalities $0 \leqslant x \leqslant 1,\ 0 \leqslant y \leqslant 1,$ $0 \leqslant z \leqslant 1 - y.$

$$\begin{aligned}
\Pr\{(X, Y, Z) \in S_1\} &= \iiint_{S_1} f(x, y, z)\, dV \\
&= \int_0^1 \int_0^1 \int_0^{1-y} (z + 2xy)\, dz\, dy\, dx \\
&= \int_0^1 \int_0^1 \left\{\tfrac{1}{2}(1-y)^2 + 2xy(1-y)\right\} dy\, dx \\
&= \tfrac{1}{3}.
\end{aligned}$$

(iii) We require

$$\int_0^1 \int_0^{1/2} \int_0^1 (z + 2xy)\, dz\, dy\, dx = \tfrac{3}{8}.$$

(iv) We in fact require $\Pr\{(X, Y, Z) \in S_2\}$ where S_2 is the tetrahedron bounded by the planes $x = 0, y = 0, z = 0, x + y + z = 1$. After some labour, we find that:

$$\iiint_{S_2} f(x, y, z)\, dV = \int_0^1 \int_0^{1-x} \int_0^{1-x-y} (z + 2xy)\, dz\, dy\, dx = \frac{7}{120}.$$

EXERCISES 7d

(1) Find the value of k such that $f(x, y) = kxy$ is a frequency function on the sample space $0 \leqslant x \leqslant 1, 0 \leqslant y \leqslant 1$. Find

(i) $\Pr\{(X, Y) \in S_1\}$, where S_1 is the rectangle given by $0 \leqslant x \leqslant \tfrac{1}{2}$, $\tfrac{1}{4} \leqslant y \leqslant 1$;

(ii) $\Pr\{X \leqslant Y\}$.

(2) Verify that $f(x, y) = 1 + 12xy$ is a frequency function on the triangle bounded by the lines $x = 0, y = 0, x + y = 1$, and find $\Pr\{(X, Y) \in S_1\}$ where S_1 is the trapezium bounded by the lines $x = 0, y = 0, y = \tfrac{1}{2}$, $x + y = 1$.

(3) Find k such that $f(x, y, z) = k\sqrt{(x^2 + y^2 + z^2)}$ is a frequency function on the sample space $x^2 + y^2 + z^2 \leqslant 1$. Find $\Pr\{Z \geqslant \tfrac{1}{2}\}$.

(4) Let f(x, y) be a frequency function defined on a sample space R. For any function g(x, y), write

$$\mathrm{E}\{\mathrm{g}(X, Y)\} = \iint_R \mathrm{g}(x, y)\,\mathrm{f}(x, y)\,\mathrm{d}A.$$

(E$\{$g$(X, Y)\}$ is called the mathematical expectation of the function g(X, Y) of random variables X and Y. A corresponding definition may be given for the mathematical expectation of a function of three random variables.)

(i) Find $\mathrm{E}\{X^r Y^s\}$ when $\mathrm{f}(x, y) = x + y$ and R is the square $0 \leqslant x \leqslant 1$, $0 \leqslant y \leqslant 1$ ($r \neq -1$ or -2, $s \neq -1$ or -2).

(ii) Find $\mathrm{E}\{X\}$, $\mathrm{E}\{Y\}$ and $\mathrm{E}\{XY\}$ when $\mathrm{f}(x, y) = (8/\pi)(x^2 + y^2)$ and R is the part of the circle $x^2 + y^2 \leqslant 1$ in the first quadrant.

(5) In the function,

$$\mathrm{f}(x, y) = 1 + \beta_1 xy + \beta_2 [x^2 - (1/12)][y^2 - (1/12)]$$

let β_1 and β_2 be constants such that $\mathrm{f}(x, y) \geqslant 0$ in the square R: $-\frac{1}{2} \leqslant x \leqslant \frac{1}{2}$, $-\frac{1}{2} \leqslant y \leqslant \frac{1}{2}$. Show that f$(x, y)$ is a frequency function on R as sample space and that

$$\mathrm{E}\{XY\} = \frac{\beta_1}{144}.$$

(6) For the frequency function $\mathrm{f}(x, y, z) = xyz$, with sample space the part of the cylinder $x^2 + y^2 \leqslant 1$, $0 \leqslant z \leqslant 4$ in the first octant, find:

(i) $\Pr\{Z \leqslant 4(X^2 + Y^2)\}$; and

(ii) $\mathrm{E}\{XYZ\}$.

Answers

With Occasional Worked Solutions

EXERCISES 1a

(i) 18, (ii) $\frac{1}{2}(\log^2 5 - 3\log^2 2)$, (iii) $\frac{3\pi}{4}$, (iv) $6\log 2 - 2$,
(v) $4 - 2\log\frac{7}{3}$.

EXERCISES 1b

(i) $\frac{2^7}{315}$, (ii) $\frac{35\pi}{2^8}$, (iii) $\frac{63\pi}{2^8}$, (iv) $\frac{16}{3003}$, (v) $\frac{3\pi}{2^7}$, (vi) $\frac{16}{1155}$,
(vii) $\frac{35\pi}{2^{16}}$, (viii) $\frac{1}{240}$.

EXERCISES 2a

(1)
(ii)

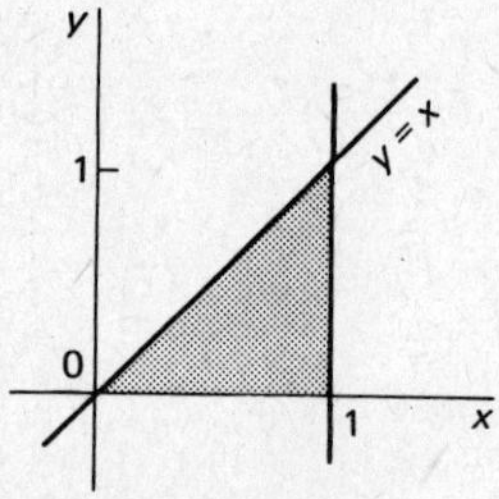

(iv)

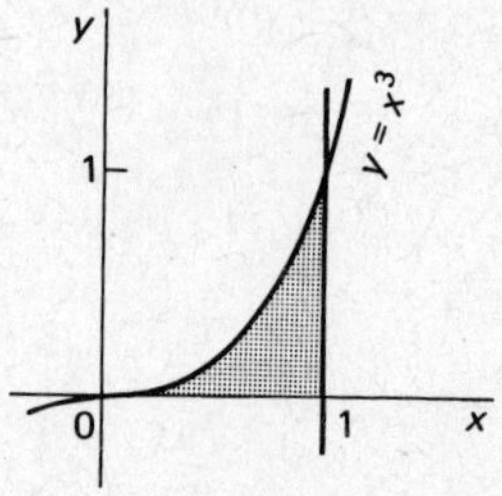

(vi)

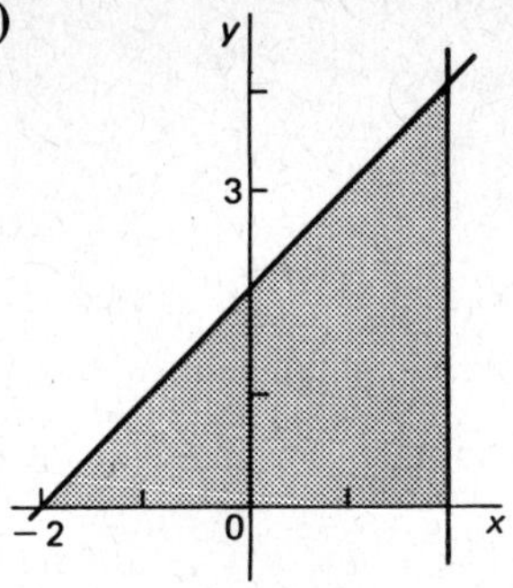

(viii)

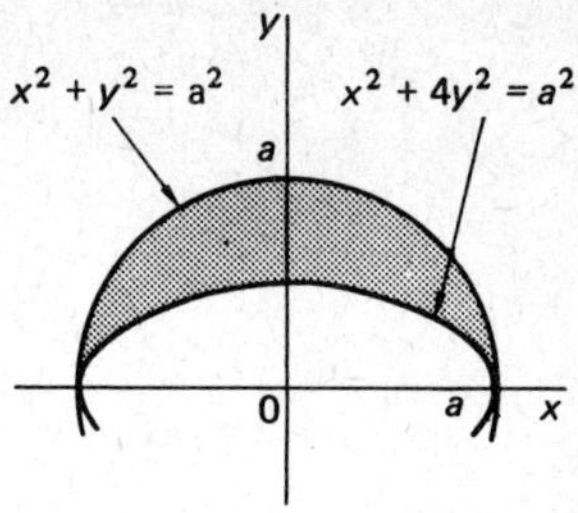

(2)

(i) $\frac{4}{3}$

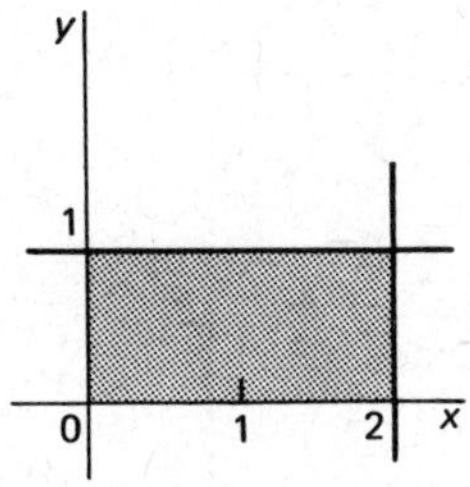

(ii) $\frac{4}{3}$

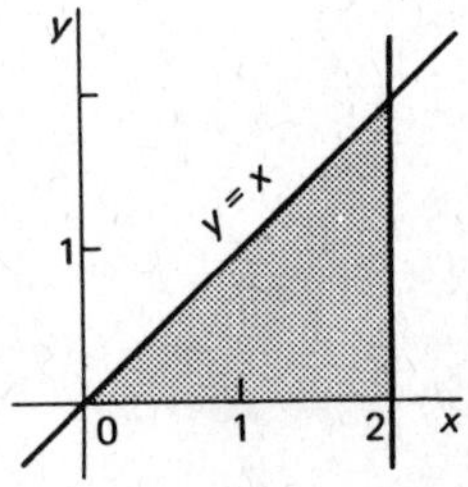

(iii) $\frac{3}{35}$

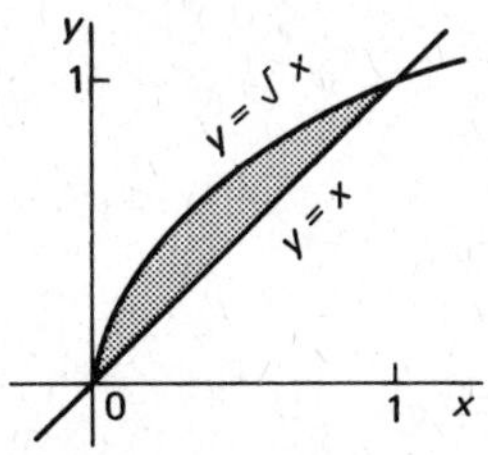

(iv) $\frac{1}{6}$

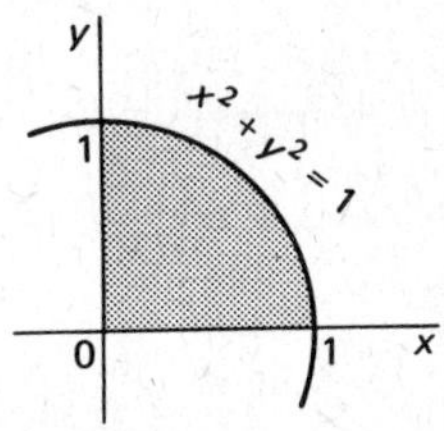

(v) $\frac{3}{5}$

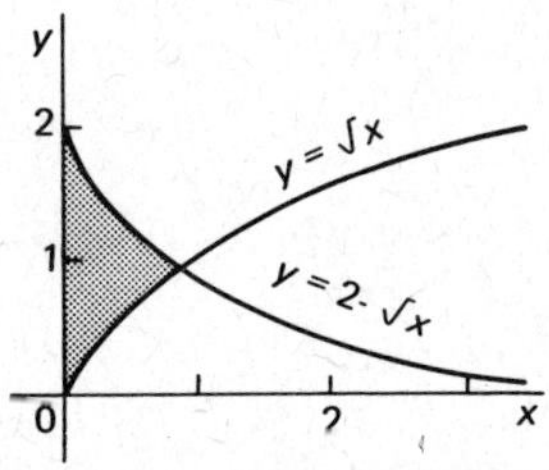

(vi) $\dfrac{a^5}{5}$, region is in (1) (viii).

(3) (i) $\frac{1}{6}$, (ii) $\frac{1}{6}$, (iii) $\frac{1}{3}$.

(4) The region is shown.

(i) $$\int_0^{\pi/2}\int_{\sin x}^{1} 2y\,\mathrm{d}y\,\mathrm{d}x = \int_0^{\pi/2} \cos^2 x\,\mathrm{d}x = \frac{\pi}{4}.$$

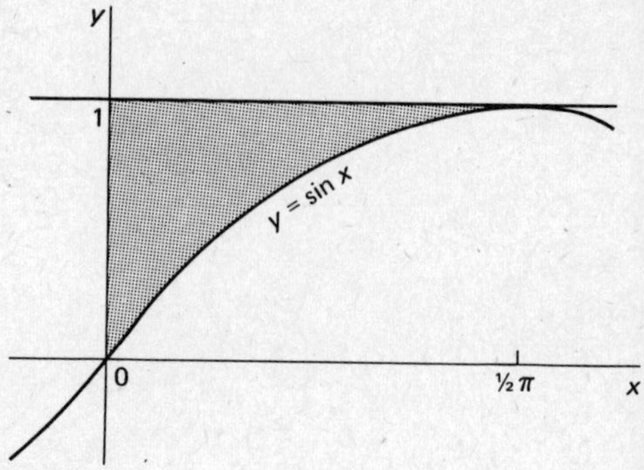

(ii) $\dfrac{\pi^2}{8}$.

(5) (i) $\dfrac{32}{3}a^3b^4$,

(ii) $$\int_0^{1/a}\int_0^{(1-ax)/b} \mathrm{e}^{ax+by}\,\mathrm{d}y\,\mathrm{d}x = \frac{1}{ab},$$

(iii) $$\int_1^2\int_{-\sqrt{(4-x^2)}}^{\sqrt{(4-x^2)}} x\,\mathrm{d}y\,\mathrm{d}x = 2\sqrt{3},$$

(iv) a^3,

(v) $$\int_0^1\int_{1-\sqrt{(1-x^2)}}^{\sqrt{(2x-x^2)}} xy\,\mathrm{d}y\,\mathrm{d}x = \int_0^1 [x^2 - x + x\sqrt{(1-x^2)}]\,\mathrm{d}x = \tfrac{1}{6}.$$

EXERCISES 2B

(1)

(i) $\frac{5\pi}{96}$

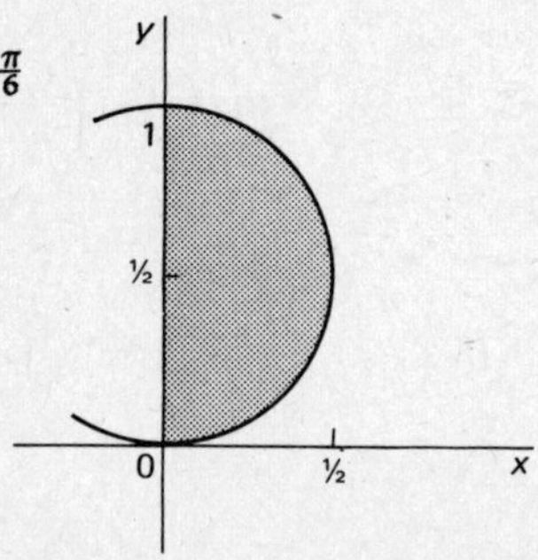

(ii) $\frac{1}{12}$

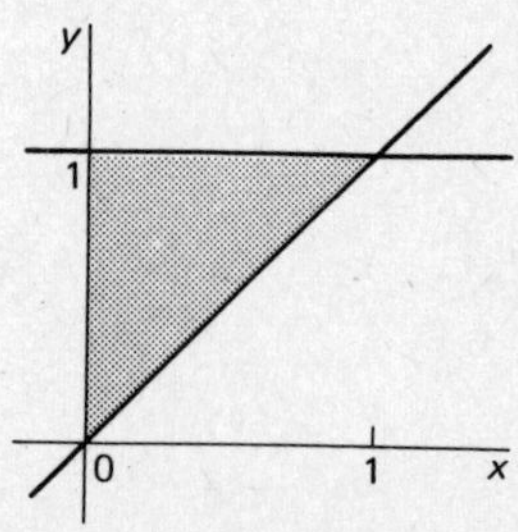

$r = \operatorname{cosec}\theta$ is $r\sin\theta = 1$ or, in Cartesian coordinates, $y = 1$.

(iii) $a^3(16-3\pi)/48$

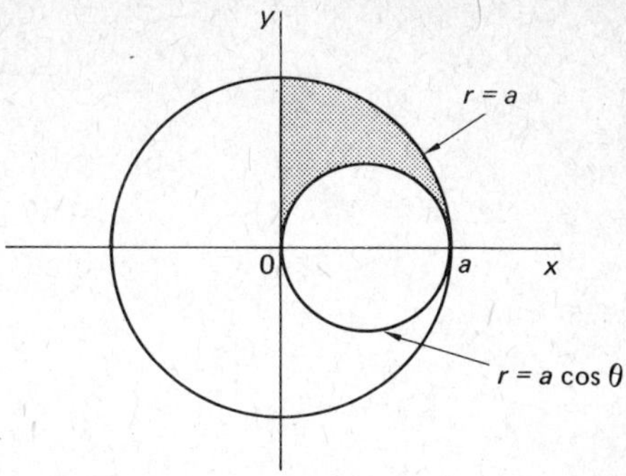

(iv) $\displaystyle\int_0^{\pi}\int_0^{a\sin 2\theta} r\sin^4\theta\,\mathrm{d}r\,\mathrm{d}\theta = \int_0^{\pi}\sin^2\theta\left[\tfrac{1}{2}r^2\right]_0^{a\sin 2\theta}\mathrm{d}\theta$

$$= \tfrac{1}{2}a^2\int_0^{\pi}\sin^4\theta\,\sin^2 2\theta\,\mathrm{d}\theta = 2a^2\int_0^{\pi}\sin^6\theta\cos^2\theta\,\mathrm{d}\theta = \frac{5\pi a^2}{64},$$

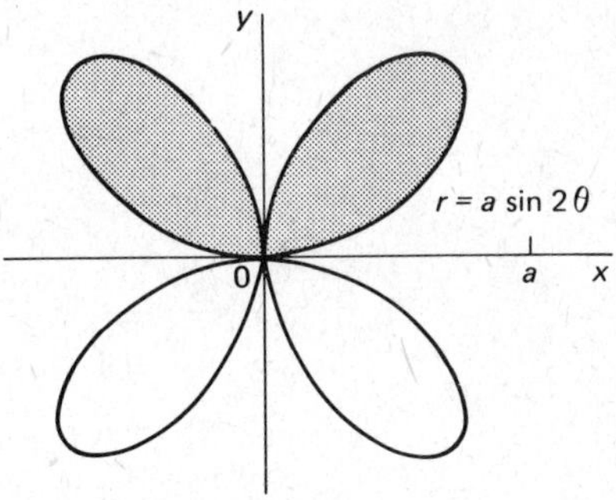

(2) (i) $\displaystyle\int_0^{\pi/2}\int_0^{a} r^4\cos^2\theta\sin\theta\,\mathrm{d}r\,\mathrm{d}\theta = \frac{a^5}{15},$

(ii) $\displaystyle\int_0^{2\pi}\int_0^{a} r^3\,\mathrm{d}r\,\mathrm{d}\theta = \tfrac{1}{2}\pi a^4,$

(iii) $\displaystyle\int_0^{\pi/4}\int_0^{\sqrt{2}} r^5\cos^2\theta\sin^2\theta\,\mathrm{d}r\,\mathrm{d}\theta = \frac{\pi}{24},$

(iv) $\displaystyle 2\int_0^{2\pi}\int_1^{b}\log r\,\mathrm{d}r\,\mathrm{d}\theta = 4\pi(b\log b - b + 1),$

(v) $\int_0^{\pi}\int_0^{a\sin\theta} r\sqrt{(a^2-r^2)}\,\mathrm{d}r\,\mathrm{d}\theta = a^3\pi/3,$

(vi) $\int_0^{\pi/4}\int_{\frac{1}{2}\sec\theta}^{\sec\theta} r^{-2}\cos\theta\,\mathrm{d}r\,\mathrm{d}\theta = \int_0^{\pi/4}\cos^2\theta\,\mathrm{d}\theta = (\pi+2)/8,$

(vii) $\frac{\pi}{2}.$

(3) (i) $\frac{2\pi a^7}{7}.$

(ii) $\int_0^{\pi/2}\int_{\cos\theta}^{1+\cos\theta} r\cos\theta\,\mathrm{d}r\,\mathrm{d}\theta = (\pi+2)/4.$

(iii) $\int_{\pi/6}^{\pi/3}\int_{\sqrt{3}}^{2\sin 2\theta} r^2\,\mathrm{d}r\,\mathrm{d}\theta = (22-3\pi\sqrt{3})/18.$

EXERCISES 3a

(1) (i) 2, (ii) $\frac{368}{27}$, (iii) $\frac{176}{5}$, (iv) $\frac{3}{28}$.

(2) (i) $\int_0^1\int_0^1\int_0^1 (x^2+2y-z)\,\mathrm{d}z\,\mathrm{d}y\,\mathrm{d}x = \frac{5}{6},$

(ii) $\int_0^4\int_0^2\int_0^2 (xz+y)\,\mathrm{d}z\,\mathrm{d}y\,\mathrm{d}x = 48,$

(iii) $\int_0^a\int_0^{\sqrt{(a^2-x^2)}}\int_0^1 y\mathrm{e}^z\,\mathrm{d}z\,\mathrm{d}y\,\mathrm{d}x = a^3(\mathrm{e}-1)/3,$

(iv) $\int_0^2\int_0^1\int_0^{x+2y+5} (x^2+y)\,\mathrm{d}z\,\mathrm{d}y\,\mathrm{d}x = \frac{82}{3}$

(v) The projection on the xy-plane is the triangle bounded by $x = 0$, $y = 0$, and $(x/a) + (y/b) = 1$;

$$\int_0^a \int_0^{b[1-(x/a)]} \int_0^{c[1-(x/a)-(y/b)]} e^{(x/a)+(y/b)+(z/c)}\, dz\, dy\, dx$$

$$= c\int_0^a \int_0^{b[1-(x/a)]} (e - e^{(x/a)+(y/b)})\, dy\, dx = bc\int_0^a (e^{x/a} - ex/a)\, dx =$$

$$= abc\,(e-2)/2.$$

(vi) The projection on the xy-plane is the triangle bounded by $x = \pi/2$, $y = 0$ and $y = x$;

$$\int_0^{\pi/2} \int_0^{x} \int_0^{x-y} \sin(x-y-z)\, dz\, dy\, dx = \frac{\pi^2}{8} - 1.$$

(3) Note that the solids may be described by the inequalities

(i) $0 \leqslant x \leqslant a, \quad 0 \leqslant y \leqslant a - x, \quad 0 \leqslant z \leqslant xy/a;$

(ii) $-a \leqslant x \leqslant 2a, \quad -a \leqslant y \leqslant 2a, \quad 0 \leqslant z \leqslant a + (x^2 + y^2)/a;$

(iii) $0 \leqslant x \leqslant a, \quad -2\sqrt{(ax)} \leqslant y \leqslant 2\sqrt{(ax)}, \quad 0 \leqslant z \leqslant a - x;$

(iv) $0 \leqslant x \leqslant a, \quad x^2/a \leqslant y \leqslant \sqrt{(ax)}, \quad 0 \leqslant z \leqslant 2a - x - y;$

(v) $0 \leqslant x \leqslant a, \quad \sqrt{(ax)} \leqslant y \leqslant 2\sqrt{(ax)}, \quad -\sqrt{(4ax - y^2)} \leqslant z \leqslant \sqrt{(4ax - y^2)}$

gives half the solid. A quarter of the required solid is shown.

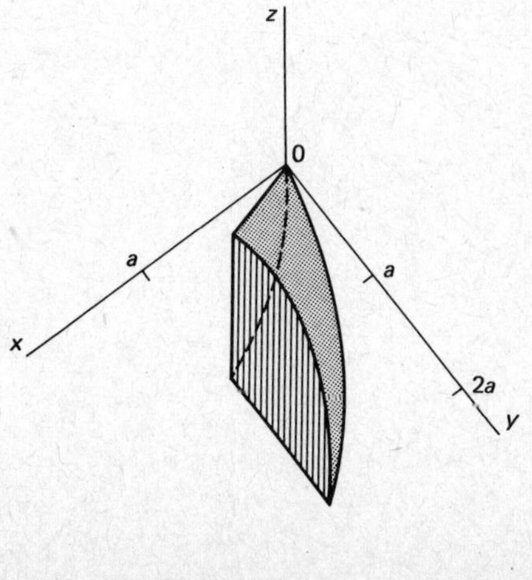

EXERCISES 3b

(1) S is described by $0 \leqslant \theta \leqslant 2\pi,\ 0 \leqslant r \leqslant a,\ 0 \leqslant z \leqslant h.$
(i) $2\pi a^3 h/3$, (ii) $\pi a^6 h^2/48$, (iii) $\pi h^3(1-\cos a^2)/3$.

(2) S is described by $0 \leqslant \theta \leqslant \pi/2,\ 1 \leqslant r \leqslant a,\ 1 \leqslant z \leqslant h.$
(i) $(a^4-1)\log h/8$, (ii) $(a^5-1)(h-1)/60$.

(3) S is described by $0 \leqslant \theta \leqslant \pi,\ 0 \leqslant r \leqslant a\sin\theta,\ 0 \leqslant z \leqslant h.$

(i) $3\pi a^4 h/32$, (ii) $\dfrac{\pi a^2 h}{12}\left(\dfrac{7a^4}{128}+h^2\right)$.

(4) Volume $= \displaystyle\int_0^{2\pi}\int_0^a\int_0^{br^2} r\,\mathrm{d}z\,\mathrm{d}r\,\mathrm{d}\theta = \pi a^4 b/2.$

(5) The wedge is described by $-\pi/2 \leqslant \theta \leqslant \pi/2,\ 0 \leqslant r \leqslant 2a\cos\theta,$ $mr\cos\theta \leqslant z \leqslant nr\cos\theta.$

(6) Volume $= \displaystyle\int_{-\pi/2}^{\pi/2}\int_0^{2a\cos\theta}\int_{-\sqrt{(2ar\cos\theta)}}^{\sqrt{(2ar\cos\theta)}} r\,\mathrm{d}z\,\mathrm{d}r\,\mathrm{d}\theta = 128a^3/15.$

(7) S is described by $-\pi/2 \leqslant \theta \leqslant \pi/2,\ 0 \leqslant r \leqslant 2a\cos\theta, -kr \leqslant z \leqslant kr.$

EXERCISES 3c

(1) (i) $\pi a^5/30$, (ii) $a^6/48$, (iii) $\pi a^2/8$.

(3) $\displaystyle\int_0^{2\pi}\int_0^{\pi}\int_b^a \frac{\sin\phi}{\rho}\,\mathrm{d}\rho\,\mathrm{d}\phi\,\mathrm{d}\theta = 4\pi\log(a/b).$

(4) $4\pi/35$.

(5) $\pi h^2(3a-h)/3$; in cylindrical coordinates,

$$\int_0^{2\pi}\int_0^{\sqrt{(2ah-h^2)}}\int_{a-h}^{\sqrt{(a^2-r^2)}} r\,\mathrm{d}z\,\mathrm{d}r\,\mathrm{d}\theta.$$

(6) The part of S in the first octant is described by $0 \leqslant \theta \leqslant \pi/2$, $\cot^{-1}k \leqslant \phi \leqslant \pi/2,\ 0 \leqslant \rho \leqslant 2a\cos\theta\,\operatorname{cosec}\phi.$

EXERCISES 4

(1) (i) $1 \leqslant y \leqslant 3,\ -2 \leqslant x \leqslant 3$; (ii) $0 \leqslant y \leqslant 1,\ y \leqslant x \leqslant 1$;

(iii) $0 \leqslant y \leqslant 1,\ \sqrt{y} \leqslant x \leqslant 1$; (iv) $0 \leqslant y \leqslant 1,\ \sqrt[3]{y} \leqslant x \leqslant 1$;

(v) $0 \leqslant y \leqslant 1,\ 0 \leqslant x \leqslant y$; (vi) $0 \leqslant y \leqslant 4,\ y-2 \leqslant x \leqslant 2$;

(vii) $0 \leqslant y \leqslant 3,\ 0 \leqslant x \leqslant 3\sqrt{y}$ and $3 \leqslant y \leqslant 6,\ 0 \leqslant x \leqslant \sqrt{(36-y^2)}$;

(viii) $0 \leqslant y \leqslant \frac{1}{2}a,\ -\sqrt{(a^2-y^2)} \leqslant x \leqslant -\sqrt{(a^2-4y^2)}$

and

$$\tfrac{1}{2}a \leqslant y \leqslant a,\quad -\sqrt{(a^2-y^2)} \leqslant x \leqslant \sqrt{(a^2-y^2)}$$

and

$$0 \leqslant y \leqslant \tfrac{1}{2}a,\quad \sqrt{(a^2-4y^2)} \leqslant x \leqslant \sqrt{(a^2-y^2)}.$$

(ii) $$\int_0^2 \int_0^x (x-y)^2\,\mathrm{d}y\,\mathrm{d}x = \int_0^2 \int_y^2 (x-y)^2\,\mathrm{d}x\,\mathrm{d}y =$$

$$= \tfrac{1}{3} \int_0^2 \Big[(x-y)^3\Big]_{x=y}^{x=2}\,\mathrm{d}y$$

$$= \tfrac{1}{3} \int_0^2 (2-y)^3\,\mathrm{d}y = 4/3,$$

(iii) $$\int_0^1 \int_{y^2}^{y} (x^2+y^2)\,\mathrm{d}x\,\mathrm{d}y,$$

(iv) $$\int_0^1 \int_0^{\sqrt{(1-y^2)}} \frac{xy}{\sqrt{(1-y^2)}}\,\mathrm{d}x\,\mathrm{d}y,$$

(v) $$\int_0^1 \int_0^{y^2} xy\,\mathrm{d}x\,\mathrm{d}y + \int_1^2 \int_0^{(y-2)^2} xy\,\mathrm{d}x\,\mathrm{d}y.$$

(5) $(\mathrm{e}-1)/2$.

(6) First show that for any function $\mathrm{f}(y)$,

$$\int_0^1 \int_0^{1-\sqrt{(1-x^2)}} x\,\mathrm{f}(y)\,\mathrm{d}y\,\mathrm{d}x = \tfrac{1}{2} \int_0^1 (1-y)^2\,\mathrm{f}(y)\,\mathrm{d}y.$$

(7)

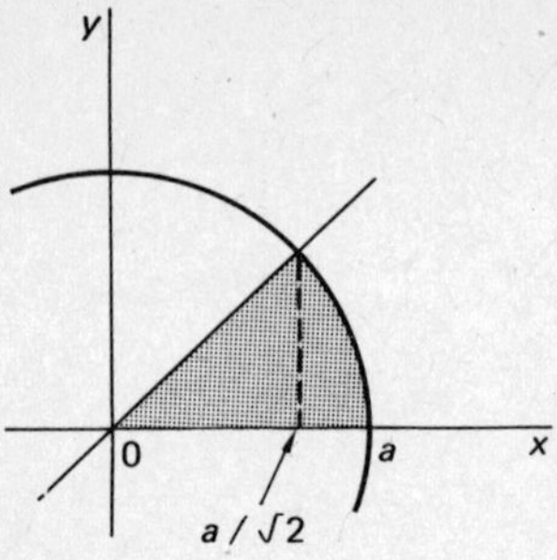

$$\int_0^{a/\sqrt{2}} \int_y^{\sqrt{(a^2-y^2)}} \cos\{c(x^2+y^2)\}\,dx\,dy$$

$$= \int_0^{\pi/4} \int_0^a r \cos c r^2\,dr\,d\theta = (\pi/8c) \sin ca^2.$$

(8) $\iiint_S \sqrt{(x^2+y^2+z^2)}\,dV = \int_0^{2\pi} \int_0^3 \int_0^z \sqrt{(r^2+z^2)}\, r\,dr\,dz\,d\theta$

$= 27\pi(2\sqrt{2}-1)/2.$

EXERCISES 5

(1) (i) The Jacobian of the transformation is 1; the circle is transformed to the interior of the circle $(u+a)^2 + (v+b)^2 = 1$, i.e. the circle of radius 1, centre $(-a, -b)$.

(ii) ab; ellipse with major and minor axes of lengths $2/a$, $2/b$, respectively.

(iii) 1; the circle is transformed into itself.

(iv) $ad - bc$; ellipse of the form $Au^2 + Buv + Cu^2 = 1$, i.e. with centre at the origin, but axes rotated.

(v) $\dfrac{-1}{(u^2+v^2)^2}$; the interior of the circle is transformed into its exterior.

(2) (i) $a^2b/3$, (ii) $\pi a^3b^3/96$, (iii) $\pi ab(a^2+b^2)/16$.

(3) Put $u = x/a$, $v = y/b$, $w = z/c$, and then use spherical coordinates.
(i) $(abc)^2/48$. (ii) $\pi abc(e-2)/2$.

(4) (ii) The integral becomes $\frac{1}{2}\int_0^1 \int_v^{2-v} u^2v^2\,du\,dv.$

(6) $a^3/7$.

(7) 2.

(8) $(\sqrt{b}-\sqrt{a})(\sqrt{d}-\sqrt{c})/2\sqrt{(abcd)}$.

(9) Put $u = y - x$, $v = y + x$. $\frac{1}{2}\int_0^1\int_{-v}^{v} e^{u/v}\,du\,dv = (e^2 - 1)/4e$.

(10) $e^4 - 4e$.

(11) $\log\left(\frac{a'}{a}\right)\log\left(\frac{b'}{b}\right)$.

(12) $2\int_\alpha^\beta\int_a^b r^2\cos 2\theta\,dr\,d\theta = (b^3 - a^3)(\sin 2\beta - \sin 2\alpha)/3$.

(13) Put $x = r\cos\theta - 2$, $y = r\sin\theta + 3$ and integrate over $0 \leqslant \theta \leqslant 2\pi$, $0 \leqslant r \leqslant 5$; $5975\pi/4$.

(14) $(4\log 2 - 1)/2$.

EXERCISES 6a

(1) (ii) If $g(x, y)$ is a continuous function in some (finite) region R which is such that $|g(x, y)| \leqslant K$ in R, then $\iint_R x^{-\alpha}y^{-\beta}g(x, y)\,dA$ converges provided both $\alpha < 1$ and $\beta < 1$. (These conditions on α and β may be relaxed if R does not contain parts of both the x- and y-axes.)

(3) (i) Exclude the origin by the line $x + y = \epsilon$ where ϵ is small and positive. The integral over the remaining trapezium is

$$\int_0^\epsilon\int_{\epsilon-x}^{1-x}\frac{dy\,dx}{(x+y)^\alpha} + \int_\epsilon^1\int_0^{1-x}\frac{dy\,dx}{(x+y)^\alpha}$$

$$= \frac{1}{1-\alpha}\int_0^\epsilon(1-\epsilon^{1-\alpha})\,dx + \frac{1}{1-\alpha}\int_\epsilon^1(1-\epsilon^{1-\alpha})\,dx$$

$$= -\frac{\epsilon^{2-\alpha}}{1-\alpha} + \frac{1}{1-\alpha} - \frac{1}{(1-\alpha)(2-\alpha)} + \frac{\epsilon^{2-\alpha}}{(1-\alpha)(2-\alpha)}$$

$$\to \frac{1}{2-\alpha} \text{ as } \epsilon \to 0,$$

provided $\alpha < 2$, $\alpha \neq 1$. If $\alpha = 1$, we obtain

$$\iint \frac{dA}{x+y} = 1 - \epsilon \to 1 \text{ as } \epsilon \to 0,$$

while if $\alpha = 2$ the integral diverges. Hence,

$$\iint_R \frac{dA}{(x+y)^\alpha} = \frac{1}{2-\alpha}, \text{ provided } \alpha < 2.$$

(ii) Exclude the line $x + y = 0$ by the line $x + y = \epsilon$, ϵ small and positive. We get:

$$\iint_R \frac{dA}{(x+y)^\alpha} = \frac{2}{1-\alpha}, \text{ provided } \alpha < 1.$$

(5) $$\int_0^M \int_y^M \frac{e^{-x}}{x}\, dx\, dy = \int_0^M \int_0^x \frac{e^{-x}}{x}\, dy\, dx = 1 - e^{-M} \to 1 \text{ as } M \to \infty.$$

(7)* See (8).

(8) Integrate over the quarter-circle $0 \leqslant \theta \leqslant \frac{\pi}{2}$, $0 \leqslant r \leqslant M$; the integral is

$$\int_0^{\pi/2} \int_0^M e^{-r^2(1 + \sin 2\theta \cos \alpha)} r\, dr\, d\theta =$$

$$= \frac{1}{2} \int_0^{\pi/2} \frac{1 - e^{-M^2(1 + \sin 2\theta \cos \alpha)}}{1 + \sin 2\theta \cos \alpha}\, d\theta$$

$$= \frac{1}{2} \int_0^{\pi/2} \frac{d\theta}{1 + \sin 2\theta \cos \alpha} - E,$$

where $E = \frac{1}{2} \int_0^{\pi/2} e^{-M^2(1 + \sin 2\theta \cos \alpha)}\, d\theta / (1 + \sin 2\theta \cos \alpha) \to 0$ as $M \to \infty$.

Then put $t = \tan \theta$ to give:

$$\tfrac{1}{2}\int_0^{\pi/2}\frac{d\theta}{1+\sin 2\theta\cos\alpha}=\tfrac{1}{2}\int_0^{\infty}\frac{dt}{\sin^2\alpha+(t+\cos\alpha)^2}$$

$$=\frac{1}{2\sin\alpha}\left(\frac{\pi}{2}-\tan^{-1}\cot\alpha\right)$$

$$=\frac{\alpha}{2\sin\alpha}.$$

When $\alpha = 0$, $\tfrac{1}{2}\displaystyle\int_0^{\infty}\frac{dt}{(1+t)^2}=\tfrac{1}{2}$, the answer to (7).

EXERCISES 6b

(1) $15\sqrt{\pi}/8$, $105\sqrt{\pi}/16$, 24, $4/15$, 2π, $2\pi/\sqrt{3}$, 2π.

(4) $2\{\Gamma(\tfrac{1}{4})\}^2/\sqrt{\pi}$.

(5) $$\frac{1}{9}\int_0^1\int_0^u v^{1/3}(1-u)^{1/2}\,dv\,du=\frac{1}{12}\int_0^1 u^{4/3}(1-u)^{1/2}\,du=\frac{1}{12}B\left(\frac{7}{3},\frac{3}{2}\right)$$

$$=2\Gamma(1/3)\,\Gamma(1/6)/935\sqrt{\pi}.$$

(6) These are particular cases of Dirichlet's Integral.

EXERCISES 7a

(2) Total surface area $=2\iint\sqrt{2}\,dA$ over the circle $x^2+y^2\leqslant 2x$, which has radius 1, and hence area π. Hence surface area is $2\sqrt{2}\pi$.

(3) We give two methods of solution.

(i) The projection of the surface on the zx-plane is the triangle $R_1: 0\leqslant x\leqslant a,\ 0\leqslant z\leqslant mx$. Write the equation of the surface as $y=\sqrt{(a^2-x^2)}$.

$$\iint_{R_1} dS=\iint_{R_1}\sqrt{\left\{1+\left(\frac{\partial y}{\partial x}\right)^2+\left(\frac{\partial y}{\partial z}\right)^2\right\}}\,dA$$

$$=\int_0^a\int_0^{mx}\frac{a}{\sqrt{(a^2-x^2)}}\,dz\,dx=ma^2.$$

(ii) The projection of the surface on the yz-plane is the quarter-ellipse $R_2: 0 \leqslant y \leqslant a, 0 \leqslant z \leqslant m\sqrt{(a^2 - y^2)}$. Write the equation of the surface as $x = \sqrt{(a^2 - y^2)}$.

$$\iint_{R_2} \mathrm{d}S = \iint_{R_2} \sqrt{\left\{1 + \left(\frac{\partial x}{\partial y}\right)^2 + \left(\frac{\partial x}{\partial z}\right)^2\right\}}\,\mathrm{d}A$$

$$= \int_0^a \int_0^{m\sqrt{(a^2-y^2)}} \frac{a}{\sqrt{(a^2 - y^2)}}\,\mathrm{d}z\,\mathrm{d}y = ma^2.$$

(5) In cylindrical coordinates, the cylinder has equation $r^2 = \cos 2\theta$.

$$\iint \mathrm{d}S = \iint \sqrt{(1 + x^2 + y^2)}\,\mathrm{d}A = 4\int_0^{\pi/4}\int_0^{\sqrt{(\cos 2\theta)}} r\sqrt{(1 + r^2)}\,\mathrm{d}r\,\mathrm{d}\theta$$

$$= \frac{4}{3}\int_0^{\pi/4} (2\sqrt{2}\cos^3\theta - 1)\,\mathrm{d}\theta = (20 - 3\pi)/9.$$

(6) Eliminating z from the two equations gives $(x + y)^2 = a^2$, so the surfaces intersect in the planes $x + y = \pm a$. Required surface area is:

$$4\int_0^a \int_0^{a-x} \frac{x + y}{\sqrt{(2xy)}}\,\mathrm{d}y\,\mathrm{d}x = 4\sqrt{2}\int_0^a \left(\sqrt{x}\sqrt{(a - x)} + \frac{(a - x)^{3/2}}{3\sqrt{x}}\right)\mathrm{d}x$$

$$= 8a^2\sqrt{2}\int_0^{\pi/2} (\sin^2\theta\cos^2\theta + \tfrac{1}{3}\cos^4\theta)\,\mathrm{d}\theta\ [x = a\sin^2\theta] = \pi a^2\sqrt{2}.$$

(7) Eliminate z from the two equations to obtain the desired projection on the xy-plane. Express the equation of this curve in polar coordinates. Then,

$$\iint \mathrm{d}S = 4\int_0^{\pi/2}\int_0^{r_0} \frac{r\,\mathrm{d}r\,\mathrm{d}\theta}{\sqrt{(c^2 - r^2)}} = 4\pi c\sqrt{(ab)},$$

where

$$r_0 = \sqrt{\left(\frac{c\cos^2\theta}{a} + \frac{c\sin^2\theta}{b} - 1\right)\Big/\left(\frac{\cos^2\theta}{2a} + \frac{\sin^2\theta}{2b}\right)}.$$

(8) The projection of the surface on the xy-plane is the region R': $x^2+y^2 \leqslant 2ax, y \geqslant 0$.

$$\iint_{R_1} dS = 4\iint_{R_1} \frac{x+y}{\sqrt{(2xy)}}\, dA$$

$$= \frac{4}{\sqrt{2}}\int_0^{\pi/2}\int_0^{2a\cos\theta} [\sqrt{(\cot\theta)} + \sqrt{(\tan\theta)}]\, r\, dr\, d\theta$$

$$= \frac{8a^2}{\sqrt{2}}\int_0^{\pi/2} (\sin^{-1/2}\theta\cos^{5/2}\theta + \sin^{1/2}\theta\cos^{3/2}\theta)\, d\theta$$

$$= \frac{2a^2}{\sqrt{2}}\left\{B\left(\frac{7}{4},\frac{1}{4}\right) + B\left(\frac{3}{4},\frac{5}{4}\right)\right\} = \frac{a^2}{\sqrt{2}}\Gamma\left(\frac{3}{4}\right)\Gamma\left(\frac{1}{4}\right) = 2\pi a^2.$$

EXERCISES 7b

(1) Let the shell have equation $b^2 \leqslant x^2+y^2+z^2 \leqslant a^2, z \geqslant 0$.
(i) $[0, 0, 3(a^4-b^4)/8(a^3-b^3)]$, (ii) $[0, 0, 15(a^6-b^6)/24(a^5-b^5)]$.

(2) $(18/35, 36/35)$.

(3) $(0, 0, 2h/3)$.

(4) (i) $\pi h(a^4-b^4)$, (ii) $\pi h(a^2-b^2)(3a^2+3b^2+4h^2)/6$.

(5) $(3/4, 3, 8/5)$.

(7) Let the spherical sector have mass M and density μ. If μ is constant,

$$M = \mu\int_0^{2\pi}\int_0^{\alpha}\int_0^{a} \rho^2\sin\phi\, d\rho\, d\phi\, d\theta = 2\pi\mu a^3(1-\cos\alpha)/3,$$

$$M\bar{z} = \mu\int_0^{2\pi}\int_0^{\alpha}\int_0^{a} \rho^3\sin\phi\cos\phi\, d\rho\, d\phi\, d\theta = \pi\mu a^4\sin^2\alpha/4.$$

Hence, $\bar{z} = 3a(1+\cos\alpha)/8$ and, by symmetry, $\bar{x} = \bar{y} = 0$. If $\mu = k\rho$, k constant,

$$M = k\int_0^{2\pi}\int_0^{\alpha}\int_0^{a} \rho^3\sin\phi\, d\rho\, d\phi\, d\theta = \pi k a^4(1-\cos\alpha)/2,$$

$$M\bar{z} = k\int_0^{2\pi}\int_0^{\alpha}\int_0^{a} \rho^4 \sin\phi \cos\phi \, d\rho \, d\phi \, d\theta = \pi k a^5 \sin\alpha/5.$$

Hence, $\bar{z} = 2a(1 + \cos\alpha)/5$ and again $\bar{x} = \bar{y} = 0$.

The surface of the cap has equation $z = \sqrt{(a^2 - x^2 - y^2)}$. Its projection on the xy-plane is the circle $x^2 + y^2 \leqslant a^2 \sin^2\alpha$.

Let the surface have density μ, constant.

$$M = \iint dS = \mu\int_0^{2\pi}\int_0^{a\sin\alpha} \frac{a\,r\,dr\,d\theta}{\sqrt{(a^2 - r^2)}} = 2\pi a^2 \mu(1 - \cos\alpha),$$

$$M\bar{z} = \iint z\,dS = \mu\int_0^{2\pi}\int_0^{a\sin\alpha} a\,r\,dr\,d\theta = \pi a^3 \mu \sin^2\alpha.$$

Hence, $\bar{z} = a(1 + \cos\alpha)/2$, and, by symmetry, $\bar{x} = \bar{y} = 0$.

(8) $(2c/3, 0, 0)$.

(9) (i) Mass is $2\pi k R^{3-n}/(3 - n)$, provided $n < 3$.

(ii) I_Z is $4\pi k R^{5-n}/3(5 - n)$, provided $n < 5$.

(10) $$M = \int_{-\pi/2}^{\pi/2}\int_0^{2a\cos\theta}\int_0^{r\cos\theta} kz\,r\,dz\,dr\,d\theta = 5\pi k a^4/8 \ (k \text{ constant}),$$

$$I_Z = 35\pi k a^6/24 = 7Ma^2/3.$$

EXERCISES 7c

(1) Proceeding as in the example,

$$V = \frac{2\pi\mu}{z}\int_{R'}^{R} \rho(\rho + z - |\rho - z|)\,d\rho, \ z > 0.$$

When $z > R$, we have $\rho < z$ so $|\rho - z| = -(\rho - z)$ and

$$V = \frac{2\pi\mu}{z}\int_{R'}^{R} 2\rho^2\,d\rho = 4\pi\mu(R^3 - R'^3)/3z.$$

When $0 < z < R'$, we have $\rho > z$ so $|\rho - z| = (\rho - z)$ and

$$V = 4\pi\mu \int_{R'}^{R} \rho \, d\rho = 2\pi\mu(R^2 - R'^2).$$

This is also true when $z = 0$.

(2) Let the sphere have equation $x^2 + y^2 + z^2 = k^2$ and let the point be $(0, 0, z)$. Let its surface density be μ, constant. The projection of the surface on the xy-plane is the circle R: $x^2 + y^2 = k^2$. Notice in the following that the upper and lower hemispheres must be considered separately. The potential is V, where

$$V = \mu \iint_R \frac{dS}{D} = \mu \iint_R \frac{1}{\sqrt{\{a^2 + b^2 + (c - z)^2\}}} \frac{k \, da \, db}{\sqrt{(k^2 - a^2 - b^2)}}$$

$$= \mu k \int_0^{2\pi} \int_0^k \frac{r \, dr \, d\theta}{\sqrt{(k^2 - r^2)}\sqrt{\{r^2 + [\sqrt{(k^2 - r^2)} - z]^2\}}}$$

$$+ \mu k \int_0^{2\pi} \int_0^k \frac{r \, dr \, d\theta}{\sqrt{(k^2 - r^2)}\sqrt{\{r^2 + [-\sqrt{(k^2 - r^2)} - z]^2\}}}$$

$$= 2\pi\mu k \int_0^k \frac{du}{\sqrt{(k^2 + z^2 - 2zu)}} + 2\pi\mu k \int_0^k \frac{du}{\sqrt{(k^2 + z^2 + 2zu)}}$$

$$[u^2 = k^2 - r^2]$$

$$= \frac{2\pi\mu k}{z}(|k + z| - |k - z|), \quad z > 0.$$

If $z > k$, $V = \dfrac{4\pi\mu k^2}{z}$; if $0 \leqslant z < k$, $V = 4\pi\mu k$.

(4) If the sphere has radius R and is centred at the origin, and if the point is $(0, 0, z)$, then the potential due to the sphere is V where

$$V = 2\pi k R^2/z \text{ if } z > R; \quad V = 2\pi k(2R - z) \text{ if } 0 \leqslant z < R \ (k \text{ constant}).$$

EXERCISES 7d

(1) $k = 4$ (i) 15/64, (ii) 1/2.

(2) 17/32.

(3) $k = 1/\pi$. $\Pr\{Z \geqslant \frac{1}{2}\} = \dfrac{1}{\pi}\displaystyle\int_0^{2\pi}\int_0^{\pi/3}\int_{1/2\cos\phi}^{1} \rho^3 \sin\phi \, d\rho \, d\phi \, d\theta = 17/96.$

(4) (i) $\dfrac{1}{(r+2)(s+1)} + \dfrac{1}{(r+1)(s+2)}$; (ii) $8/5\pi$, $8/5\pi$, $2/3\pi$.

(6) (i) $\displaystyle\int_0^{\pi/2}\int_0^1\int_0^{4r^2} r^3 z \sin\theta \cos\theta \, dz \, dr \, d\theta = 1/2,$

(ii) $\displaystyle\int_0^{\pi/2}\int_0^1\int_0^4 r^5 z^2 \sin^2\theta \cos^2\theta \, dz \, dr \, d\theta = 4\pi/18.$

Index